'This book manages both to advance the academic understanding of emerging phenomena of governance of climate adaptation as well as to significantly enrich and develop the debate about action-research. By providing practical experiences from a highly innovative Dutch research programme and contrasting them with case studies from other countries, the volume helps not only academics, but also societal decision makers to address the manifold challenges of climate adaptation more effectively.'

Bernd Siebenhuber, Carl von Ossietzky University of Oldenburg, Germany

'Given the failure of collective action on global carbon emissions reductions, collective action to manage the consequences of climate change becomes one of the great challenges of the future. This book is an important contribution to understanding key problems in governing climate adaptation, drawing on over ten years of Dutch and international research.'

Frans Berkhout, King's College London and Future Earth, UK

First published 2015 by Routledge

2 Park Square, Milton Park, Abingdon, Oxon OX14 4RN
711 Third Avenue, New York, NY 10017, USA

Routledge is an imprint of the Taylor & Francis Group, an informa business

First issued in paperback 2016

British Library Cataloguing-in-Publication Data
A catalogue record for this book is available from the British Library

Library of Congress Cataloging-in-Publication Data
A catalog record for this book has been requested

ISBN: 978-1-138-01760-3 (hbk)
ISBN: 978-1-138-28232-2 (pbk)

Typeset in Goudy
by HWA Text and Data Management, London

Action Research for Climate Change Adaptation

Developing and applying knowledge for governance

Edited by Arwin van Buuren,
Jasper Eshuis and Mathijs van Vliet

Routledge
Taylor & Francis Group

LONDON AND NEW YORK

Contents

Figures

Tables

Contributors

Daan Boezeman is a PhD candidate at the Radboud University Nijmegen, the Netherlands. His dissertation research deals with knowledge production for climate adaptation in different science-policy interfaces. His main research interests are science and technology studies, institutional perspectives and the role of ideas in the policy process.

Bas Breman works as a researcher, project manager and research coordinator at Alterra, a research institute of Wageningen University and Research Centre, the Netherlands. His expertise lies at the intersection of social and spatial development, both in the field of water management and rural development.

Arwin van Buuren is Associate Professor of Public Administration at the Erasmus University Rotterdam, the Netherlands. His research focuses on the governance of climate change adaptation and flood risk management.

Art Dewulf is Associate Professor at the Public Administration and Policy Group at Wageningen University, the Netherlands. He has been involved in research projects on river basin management, adaptive water management and climate adaptation governance.

Patrick Driscoll is a PhD candidate within the Department of Development and Planning at Aalborg University, Copenhagen, Denmark. His primary research focus is on the socio-cognitive aspects of decision-making within climate change planning, including heuristics, biases, and framing effects.

Mike Duijn is senior researcher with the department of Public Administration of the Erasmus University Rotterdam and with the Netherlands Organisation for Applied Scientific Research (TNO). He specializes in applying qualitative methods for participatory research and policy making.

Gerald Jan Ellen is senior researcher/consultant public administration at Deltares, the Netherlands, where his work is organized around the concept of deltagovernance: an integral governance approach on water, soil and spatial planning.

Jasper Eshuis is Assistant Professor of Public Administration at the Erasmus University Rotterdam, the Netherlands. His research is on the governance of complex systems, with special interests in spatial planning, as well as public branding and place marketing.

Patrick Huntjens is Head of Water Diplomacy at The Hague Institute for Global Justice, the Netherlands, where he is responsible for several international projects on water diplomacy and water governance.

Wiebren Kuindersma is a policy scientist for Alterra, a research institute of Wageningen University and Research Centre, the Netherlands, working on policy evaluations and other policy studies on environmental and spatial issues. He is also working on his PhD thesis at the Forest and Nature Conservation Policy Group at Wageningen University, the Netherlands.

Frank van Lamoen is a geographer who works on climate change adaptation and water management for the Province of North Brabant, the Netherlands. He coordinates research activities for the dry rural areas in the Netherlands for the national Knowledge for Climate programme and he is leader of a working group on water governance within the European Innovation Partnership Water.

Ralph Lasage is researcher at the Institute for Environmental Studies at the VU University Amsterdam, the Netherlands. His work is on changes in the hydrological cycle in river basins due to climate change and land-cover change, how this affects people and ecosystems, and how adaptive measures can help meliorating the impacts.

Corniel van Leeuwen is junior researcher and PhD candidate at the Department of Public Administration (Faculty of Social Sciences) at the Erasmus University Rotterdam, the Netherlands, in close co-operation with Deltares. His research focuses on the effectivity of methods and instruments used by practitioners to connect time spans in complex decision making processes related to climate adaptation and water governance.

Martin Lehmann is Associate Professor of Sustainable Development at the Department of Development and Planning, Aalborg University, Denmark. He is also co-founder of KlimaLab, a climate change innovation laboratory aimed at rapidly scaling climate action and solutions, and the co-founder and deputy head of The Danish Centre for Environmental Assessment.

Pieter Leroy is Professor of Political Sciences of the Environment at the Nijmegen School of Management, Radboud University Nijmegen, the Netherlands. His research focuses on institutional analysis in environmental policy, in particular on the emergence and functioning of new policy arrangements, in a context of more encompassing societal and political changes.

John Martin is Professor at La Trobe University and Director of the Centre for Sustainable Regional Communities based in Bendigo, Australia. His research focuses on the sustainability of rural towns and regional communities.

Bouke Ottow has worked as a professional in water management for 30 years and was involved in participation processes as a designer and facilitator in the major river basins in the Netherlands, as well as in Indonesia, Vietnam, the Philippines, Romania, Egypt and Kenya.

Rob Roggema is a landscape architect and an internationally renowned design expert on climate adaptation, low carbon living and sustainable development. He currently holds professorships at Van Hall Larenstein University for Applied Sciences in Velp and Inholland University of Applied Sciences in Delft, the Netherlands, as well as the National Institute for Design Research, Swinburne University of Technology, Australia.

Todd Schenk is the Assistant Director of the MIT Science Impact Collaborative and a PhD candidate in the Environmental Policy and Planning Group of the Department of Urban Studies and Planning at the Massachusetts Institute of Technology, USA. His work focuses on how planners, decision-makers and other stakeholders can collaboratively make effective decisions in science-intensive situations that involve complex risks and high degrees of uncertainty.

Lawrence Susskind is the Ford Professor of Urban and Environmental Planning at the Massachusetts Institute of Technology, Director of the MIT Science Impact Collaborative, Vice Chair of the Program on Negotiation at Harvard Law School, and Founder and Chief Knowledge Officer at the Consensus Building Institute, USA.

Catrien Termeer is Professor of Public Administration and Policy at Wageningen University, the Netherlands. Her research addresses the governance of wicked problems in the domains of climate, water, and agriculture. She studies innovative approaches of governance, varying from self-governing communities and collaborative practices to global round tables and adaptive management systems. Her research is both scientific relevant and useful and recognizable to practitioners.

Ellen Tromp is senior researcher at the research institute Deltares, the Netherlands. She holds master's degrees in public administration (Erasmus University Rotterdam) and in civil engineering (University of Technology, Delft). Her research focuses on urban dyke design and multi-purpose levees. Her work is organized around the concept of deltagovernance: an integral governance approach on water, soil and spatial planning.

Peter van Veelen is a senior urban planner at the City of Rotterdam, the Netherlands, department of Urban Design, Planning and Housing. He is responsible for the development of a climate-adaptive flood risk strategy for the flood-prone areas of the city of Rotterdam. As a part-time PhD researcher at the TU Delft, Department of Urbanism, he is working on developing spatial concepts and planning strategies of integrating flood risk strategies with urban development processes.

Martinus Vink is PhD candidate in the field of public administration at Wageningen University, the Netherlands. In his research he focuses on the role of framing and frame interactions in governance processes where actors simultaneously puzzle over long term climate change impacts and power over short term priorities. Previously he worked for the Dutch Scientific Council for Government Policy (WRR) where he contributed to various scientific reports to the Dutch government on long-term policy issues. He has published work in journals such as *Environmental Science & Policy*, *Ecology and Society*, and *Journal of Comparative Policy Analysis*.

Mathijs van Vliet is postdoctoral research fellow of public administration at Wageningen University, the Netherlands. He works on action research within the Governance of Climate Adaptation Consortium in the Knowledge for Climate Research Programme. He recently organized an extensive co-creation project in which high-level policy makers collaborated with senior scientist to develop practical guidelines for the governance of adaptation of climate change. His research focusses on water, spatial planning, climate change and scenario development.

Lisa Vos has 20 years of experience in the fields of leadership development, organisational development and executive education. She holds an MSc in business administration (Rijksuniversity of Groningen, The Netherlands, 1995). She was, amongst others, Senior Consultant Executive Education at Melbourne Business School and Program Manager Executive Programs at Nyenrode Business University, where she designed and delivered leadership development programs (both open and customized) for senior managers from a broad range of organizations. Lisa currently works from her private consultancy practice.

Preface

Writing a book on the governance of climate change adaptation is an exciting endeavour. Climate change is a complex and controversial issue. Adaptation to climate change is even more intricate as it has to deal with the physical complexities and the social controversies of climate change, as well as the complex governance setting. This book aims to contribute to this exciting field.

Given the complexities of adaptation to climate change, there is a need for scientific knowledge. More specifically, there is a need for scientific knowledge that is applicable in concrete governance practices. This book is about doing action research in the field to develop such knowledge; and everyone involved in doing action research knows that it is really challenging to act on the edge of science and practice. Especially when this practice is about such a wicked issue as the governance of climate change adaptation!

It is thus almost unnecessary to mention that writing this particular book was an unforgettable experience. This book emerged out of the Knowledge for Climate Programme on Governance of Adaptation. In this programme, action research was chosen as a common methodological starting point. During the project, the idea was born of comparing our experiences with those of other countries. The European Climate Change Adaptation Conference in Hamburg (2013) gave the final push to writing this book.

Our pleasure in doing this was greatly enhanced by the enthusiastic and supportive involvement of the authors involved. Not only were they willing to apply our rather strict guidelines on how to structure their chapter and describe their action research project, they also selflessly cooperated in an intensive author workshop to think about the aim of the book and its main conclusions.

Second, we would like to thank the Routledge team who accompanied the publication of our book: Helen Bell, Louisa Earls, and Bethany Wright. They were genuinely supportive during the whole process from our initial idea to delivery of the final manuscript. Without their cooperation and encouragement, this book would never have been produced.

Many thanks are due to Catherine O'Dea for editing all the chapters, checking the references, and applying a consistent style. The same holds true for Jitske Verkerk, who assisted us in the final stage of our book, managing the timely and correct delivery of the manuscript.

Without the financial support of the Knowledge for Climate Foundation, this book could not have been written; and without the financial support of Erasmus University Rotterdam and Wageningen University, the book could not have been published. We appreciate this support very much.

Arwin van Buuren
Jasper Eshuis
Mathijs van Vliet

1 The governance of adaptation to climate change and the need for actionable knowledge

The challenges of climate change adaptation and the promise of action research

Arwin van Buuren, Mathijs van Vliet and Catrien Termeer

Introduction

All over the world, governments are considering how to deal with the issue of climate change and its possible consequences. Most energy is devoted to the question of how to mitigate climate change by reducing CO_2 emissions. However, there is growing evidence that climate change will be irreversible and will have serious threatening consequences. Therefore, the issue of how we can and have to adapt to changing climate conditions is coming more and more onto the political and administrative agenda (IPCC, 2012). The taboo that for a long time rested on adaptation is slowly being lifted (Pielke *et al.*, 2007).

The policy domain of climate change adaptation is thus relatively new; this means that policy ambitions are still under construction, and the same holds for policy instruments and arrangements. In many countries, governments are still busy with exploring the possible impact of climate change, whereas in other countries authorities are drafting ambitious programmes to make society climate-proof. Many authors urge for more decisive action and are anxious about the indecisiveness that they are witnessing (Giddens, 2009; Hulme, 2009).

The fact that climate change adaptation is still in its infancy also implies that there is a strong perceived need for policy-relevant information. Much time, money, and effort is invested in conducting technical analyses about what climate change can and does imply, which policy options are available, and how effective they are (Bruin *et al.*, 2009). There is, however, also a growing need for information on how to govern climate change adaptation and deal with the specific challenges it poses; in other words, how to *govern adaptation to climate change* (Termeer *et al.*, 2011). In particular, questions like how to mainstream adaptation into other policy arenas, how to mobilize private actors to finance adaptation measures, and how to connect short-term and long-term policy ambitions are important for many policymakers at multiple governance levels. Other questions relate to how to design institutional arrangements that fit the adaptation ambition (Hallegatte, 2009).

For social scientists, it is interesting to see how the climate change adaptation domain is evolving, the frames used to put climate change on the agenda, the barriers met, and the mix of instruments chosen (Biesbroek *et al.*, 2010; Hulme, 2009; Pralle, 2009). However, because of a serious lack of systematic, comparative, and evaluative research, it is difficult to translate this emerging body of knowledge into policy-relevant insights. Part of the problem relates to a lack of knowledge of suitable research methods that fit into the emerging policy adaptation domain – which is characterized by much complexity and uncertainties – and that result in both scientifically sound and application-oriented knowledge.

In this book, we explore the potential for action research. We define action research rather broadly as a research methodology in which researchers enter real-world situations and aim both to improve it and to acquire knowledge (Checkland and Holwell, 1998). Action research approaches share the aim of building 'theories within the practice context itself and test them through intervention experiments' (Argyris and Schön, 1991: 86). Doing scientific research *in situ* seems to be highly relevant in a context in which much has to be invented. Researchers are involved in a much more active way compared to the traditional observant role, on the one hand giving them the opportunity to get detailed empirical insight information that they might not obtain with traditional research methods, and on the other urging them to design interventions that improve the practices that they are analysing.

This book presents and analyses a variety of action research practices in the field of *governance of climate change adaptation* and reflects on the current action research literature. Case studies from diverse countries show how action research works in this complex and relatively new field. Special attention is paid to the potentials and pitfalls of action-oriented research approaches. At the same time, the case studies unveil new insights and different practices of governance of adaptation to climate change. The book has both a methodological and a prescriptive aim. First, the book explores the different methods of action research in the field of governance of adaptation to climate change and analyses how action research methods are being applied in this complex and relatively new context, and the potentials and pitfalls faced by researchers. Second, the book has a prescriptive ambition because it contributes to the development and effective application of action-oriented research approaches in the domain of climate change adaptation, through learning across cases, places, and methods.

In this chapter, we substantiate our argument that the status of climate change adaptation and the characteristics of this emerging domain are fertile ground for action-oriented research approaches. In the next section, we show that climate change adaptation is as much a governance challenge as a technical issue. In the third section, we argue that, because it is an immature policy domain and because policymakers are faced with huge uncertainties and controversies, a more collaborative interaction between social scientists and policymakers or planners could be helpful in realizing more effective governance strategies. Then we describe in more depth the expected benefits of action research in the context of climate change adaptation. Thereafter, we formulate the key questions to be answered in this book. The final section gives an overview of the remaining chapters.

Climate change adaptation as a governance challenge

There is increasing recognition of the need for society to adapt to the impacts of climate change (IPCC, 2012). Climate change adaptation involves technical adjustments, like raising dykes or creating water storage, but also calls for broader processes of societal change and transitions, and for increasing the adaptive capacity of society to deal with unexpected future changes (Jordan *et al.*, 2010). The governance of adaption will face all the usual difficulties, hindrances, and opportunities of dealing with complex problems. On top of that, adaptation to climate change poses some specific, particularly demanding, governance challenges and dilemmas (see e.g. Haug *et al.*, 2010; Termeer *et al.*, 2011).

The governance of adaptation: challenges

A number of governance challenges characterize the field of climate change adaptation. First, a multi-actor, multi-sector, and multi-level governance world forms the inescapable context for climate change adaptation, because the ramifications of climate change stretch across different policy domains and institutional levels. Adaptation is highly interconnected, stretching over policy fields as varied as water management, spatial planning, infrastructure, agriculture, energy supply, industry, nature, and health. Climate change potentially impacts upon all these fields, and the interactions between them. Within each field, there are also increasingly complex governance systems that involve not only governmental actors, but also businesses and other civil society actors, at local, regional, and national level. Successful adaptation is highly dependent upon the ability to mainstream adaptation with other – existing – policy domains (Uittenbroek *et al.*, 2013) and also upon the involvement and collaboration of many actors from these fields, with their own ambitions and preferences, responsibilities, problem definitions, and resources (see Boezeman *et al.*, Chapter 5, this book). Governance strategies need to deal with this fragmentation (Verkerk *et al.*, forthcoming). Taking adaptation measures thus also leads to complex coordination issues and institutional flurry. These characteristics make it far from easy to formulate legitimate adaptation strategies (van Buuren *et al.*, 2014) and implement them (see Ellen *et al.*, Chapter 7, this book). In spite of these inherent uncertainties and ambiguities, decisions about adaptation strategies need to be taken or prepared now. The fragmented context of adaptation leads to many ambiguities when it comes to the question of who is responsible for what (Brouwer et al,. 2013).

Second, climate change adaptation is still in its infancy and lacks a well-structured policy domain and practice. This further increases the ambiguity about rules, roles, and responsibilities in the adaptation domain. Climate impacts will not affect all sectors and actors in the same way. It is therefore necessary to deal with the distribution of risks, costs, and benefits. Moreover, because of the multi-sectoral nature of climate change adaptation, redistribution of responsibilities is also needed. Land-use planners will have to deal with water-management issues, and water managers have to take into account the threats of new insects with new diseases, for example.

Third, decision making in relation to climate change is knowledge intensive, and important uncertainties about the nature and scale of risks and about the effectiveness of solutions will persist (Termeer *et al.*, 2011). There are still important uncertainties about the impacts of climate change and the effectiveness of adaptation measures (Arvai *et al.*, 2006). In addition, because climate change is controversial, climate change adaptation is controversial too (see Vink *et al.*, Chapter 3, this book). Organizing enough support for adaptation to uncertain climate changes is thus far from easy. Controversy is inevitable when the many actors involved bring with them a variety of frames to make sense of a high-stake issue like climate change (Dewulf, 2013; Hulme, 2009). Differences in frames and perspectives affect not only the interpretation of knowledge, but also the desirability of adaptation options and connected governance arrangements (see van Buuren *et al.*, Chapter 10, this book). The redefinition of rights and obligations further contributes to the controversy.

Finally, the consequences of climate change manifest themselves in the future; this gives decision makers the time to implement adaptation measures, but they have to be drafted without absolute certainty about the consequences (see Huntjens *et al.*, Chapter 4, this book). This often leads to a delay in decision making (Fankhauser *et al.*, 1999; Hulme, 2009). We need to find ways to link long-term problems to present-day solutions and develop them in a robust way so as to deal with part of the long-term uncertainty (van Leeuwen and van Buuren, 2013).

The governance of adaptation: international examples

Notwithstanding this complex character of adaptation to climate change, policymakers all over the world have started to develop more or less ambitious adaptation programmes. However, both the scope and intensity of national programmes varies significantly (Table 1.1). In this section, we give a variety of examples which give a first impression of how various countries deal with adaptation to climate change.[1]

There are significant differences regarding the scope of national adaptation strategies. It is quite surprising that there is a strong bias towards domains that are traditionally seen as vulnerable to natural disasters, whereas much less attention is paid to things like heat stress, green infrastructure, industry, and natural disasters.

With regard to the governance of climate change adaptation, there is a strong focus on flood management and other water-related issues. Furthermore, there is a growing awareness of the importance of mainstreaming adaptation into other policy domains (Uittenbroek *et al.*, 2013).

Table 1.1 Policy sectors addressed in European adaptation strategies (EEA, 2013: 75)

Sectors	Number of countries	AT	BE	BG	CH	CY	CZ	DE	DK	ES	FI	FR	GR	HU	IE	IT	NL	NO	LT	LV	PL	PT	RO	SK	SI	UK
Water management and water resources	23	x	x	x	x	x	x	x	x	x	x	x	x	x	x	x	x	x			x	x	x	x	x	x
Forests and forestry	23	x	x	x	x	x	x	x	x	x	x	x	x	x	x	x	x	x			x	x	x	x	x	x
Agriculture	22	x	x	x	x	x	x	x	x	x	x	x	x	x	x	x	x			x	x	x	x	x	x	
Biodiversity, ecosystem services	19	x	x		x	x	x	x	x	x	x	x	x		x	x		x			x	x		x		x
Human health and wellbeing	18	x	x	x	x	x	x	x	x	x	x	x	x		x	x	x		x	x	x	x				x
Infrastructure and built environment	14	x					x	x	x	x	x	x		x	x	x	x	x			x					x
Spatial planning, urban planning, and development	14	x		x		x	x		x	x		x	x	x		x				x	x	x				
Energy, energy consumption	14	x				x	x		x	x	x	x	x		x	x			x		x	x	x			x
Coastal areas, coastal management	13		x			x		x	x	x	x	x	x		x	x			x	x	x	x				
Tourism	13	x		x			x		x	x	x	x	x			x					x	x				
Civil protection, safety, preparedness, and rescue services	10	x					x		x			x	x			x	x			x		x				x
Transport, transport infrastructure	10	x							x	x	x	x				x		x	x		x					x
Fishery and aquaculture	9			x		x			x	x	x	x			x	x						x				
Industry	8								x	x	x	x			x	x						x				
Natural disasters/hazards	5	x			x					x		x	x													
Soils and desertification	5			x				x		x		x	x							x						
Business and services	5							x	x	x	x									x						
Green infrastructure, urban green spaces	2	x												x												
Economy	2	x																		x						
Regional development	2								x					x												
Communities	2										x															
Heat-related issues	1		x																							
Mountain areas	1									x																

In a flood-prone country like the Netherlands, climate change adaptation is highly focused upon flood risk safety. In 2009, the Dutch Delta Programme started with the aim of taking long-term decisions to keep the Dutch Delta safe from floods, provide it with enough fresh water, and anticipate climate change. It is based on the recommendations of the 2008 Delta Committee, which recommended a Delta Act that should constitute the basis of a Delta Programme, a Delta Fund, and a Delta Commissioner to institutionally bridge the usual four-year policy cycles and guarantee the long-term character of the climate change adaptation measures when national budgets become tight. The Delta Act came into force in 2012 (Boezeman *et al.*, 2013; Vink *et al.*, 2013).

The Delta Programme consists of three thematic sub-divisions (water safety, fresh water, and urban and spatial planning) and six geographical sub-divisions. The programme is led by a Delta Commissioner, who serves as a liaison between the government (local authorities, water boards, provincial authorities, and ministries), civil society organizations, and other stakeholders. Many meetings have been held and studies have been done. The Commissioner is responsible for combining the insights from the different sub-programmes and will present the Delta Decisions in September 2014 (Termeer *et al.*, forthcoming). Despite the Delta Programme's efforts, the Dutch Audit Council assessed the Dutch adaptation strategy as too narrow and too fragmented. With the Climate Agenda (Ministerie van Infrastructuur en Milieu, 2013), Dutch politics responded by announcing climate risk assessments for vulnerable sectors such as energy, public health, infrastructure, and nature, in order to draft a more comprehensive national adaptation programme before 2017.

At European level, the governance of climate change adaptation is approached from a more integrated perspective. In April 2013, the European Commission (EC) presented its strategy on adaptation to climate change (EC, 2013). The strategy is accompanied by documents on adaptation in specific sectors and policy areas, such as migration, marine and coastal areas, health, infrastructure, agriculture, cohesion policy, and insurance. It further includes guidelines for the member states (MSs) on preparing national adaptation strategies (EC, 2013). The strategy's overall aim to contribute to a more climate-resilient Europe (EC, 2013) splits up into three goals supported by eight actions:

1 promote and support MSs to develop national adaptation strategies and take concrete actions via the provision of guidelines and funding to support capacity building;
2 ensure better-informed decision making by filling knowledge gaps on adaptation costs and benefits, risk assessments, decision support models, tools and frameworks, monitoring and evaluation methods, as well as further developing the CLIMATE-ADAPT web portal (http://climate-adapt.eea.europa.eu/); and
3 climate-proofing EU action via mainstreaming adaptation into EU policies and programmes. This has already been done for the sectors mentioned

above; in the near future, other policies like the Common Agricultural Policy, Common Fisheries Policy, and Cohesion Policy will follow.

The EU cannot force member states to take action and develop national adaptation strategies, as it has no mandate in this field. This is a major reason for emphasizing mainstreaming climate change adaptation into EU initiatives, such as Europe's growth plans and sectors in which it does have the power to force member states to act (EEA, 2013). In the 2014–2020 budget, 20 per cent of disbursements should be climate related (EC, 2013). Multiple adaptation projects have already received funding from, for instance, the European Regional Development Fund (EEA, 2013).

The EC realizes that most of the actual adaptation should be done at local and regional level and that climate change will have different effects in the different MSs. Yet, it sees a role for itself, as a lack of adaptation in one MS might negatively affect neighbouring countries (EC, 2013). Moreover, MSs can learn from one another, and the EU can assist in bridging knowledge gaps and capacity building (Termeer *et al.*, forthcoming).

On the global scale, the UN Framework Convention on Climate Change (UNFCC) aims to prevent 'dangerous' human interference with the climate system. From its inception in 1992, the focus lay mainly on mitigation, but in 2001 three funds were set up to support adaptation, among which is the Adaptation Fund funded via the Clean Development Mechanism of the Kyoto Protocol. Also, the Green Climate Fund should provide developing countries with funding for adaptation. However, little progress has been made, and available funding is inadequate to meet even the most urgent needs of developing countries (Verschuuren, 2013). In 2010, the Cancun Adaption Framework was adopted, requiring countries to plan, prioritize, and implement adaptation actions and strengthen institutional capacities (Termeer *et al.*, forthcoming).

Knowledge for adaptation: the need for reflexive and application-oriented research

In this complex context and with such a complex issue at stake, policymakers can benefit from insights from social sciences like economics, sociology, law, public administration, and political science when drafting adaptation strategies. There are many examples of large-scale research programmes aiming to deliver policy-relevant knowledge, to make adaptation policies more evidence based, and to deliver usable insights about governance arrangements, procedures, and strategies. Again, a short overview can give an impression of what is happening in this field.

In Germany, the Federal Ministry of Education and Research (BMBF) is funding KLIMZUG – Managing Climate Change in the Regions for the Future – to stimulate the development of innovative approaches to climate change adaptation. It contains a number of projects spread over Germany, with a strong focus on network development and interaction, capacity building, and institutional development (http://www.klimzug.de/en/160.php).

Adaptation in France is rather diverse, even though the focus is predominately on energy mitigation. Under the Sarkozy administration, France started to organize adaptation in a rather hierarchical way, for instance with a law that forces communities to make a climate plan (Grenelle II Act: http://www.developpement-durable.gouv.fr/IMG/pdf/Grenelle_Loi-2.pdf). In general, the boundaries between science and policy in France are rather strict. In 2008, club ViTeCC (an expert network organization) was founded to assist cities and local administrations with their adaptation efforts. This boundary organization is a collaboration of different knowledge institutes, develops local impact studies, and functions as a platform to bundle expertise. Several other experiments have also been undertaken, for instance the setting up of a regional IPCC for the Bordeaux region.

Go-Adapt is an Austrian political science research project that studies the governance of climate change adaptation (http://www.wiso.boku.ac.at/go-adapt.html). It studies three governance challenges perceived as important in the context of climate change adaptation: improving horizontal and vertical policy integration, coping with uncertainties, and stakeholder involvement. It aims, among other things, to provide guidance on the establishment of climate change adaptation policy frameworks and thus has a strong focus on delivering application-oriented knowledge.

In the Netherlands, the Knowledge for Climate (KfC) programme, a large-scale scientific programme, ran from 2011 to 2014. It was preceded by the Climate Changes Spatial Planning research programme. Within the KfC, the consortium on the governance of adaptation to climate change tried to apply action research on a large scale. Three examples of this work are given in this book, dealing with flood risk management and fresh water availability.

At European level, numerous climate change projects have been taking place. JPI Climate is a collaboration between 13 European countries to coordinate jointly their climate research and fund new transnational research initiatives (http://www.jpi-climate.eu/home). CIRCLE-2 is a European network of 34 institutions from 23 countries committed to funding research and sharing knowledge on climate change adaptation and the promotion of long-term cooperation among national and regional climate change programmes (http://www.circle-era.eu/np4/home.html).

On the global scale, the World Bank is financing research projects on climate change (http://www.worldbank.org/en/topic/climatechange/projects), for instance in Kenya, Chile, and Laos. The 2007 IPCC report also includes information on adaptation to climate change, including an assessment of adaptation practices, options, constraints, and capacity (Adger *et al.*, 2007).

These examples of application-oriented research bring us to the question of how scientific knowledge can contribute to the governance of adaptation to climate change. After all, there are many problems when it comes to bridging the gap between knowledge and policy. Often, traditional research programmes fall short of becoming relevant and making the step from pure science towards utilization and application. Here, we come to our argument that other research approaches are necessary to prevent misfits between policymaking and research. Especially in the emerging domain of adaptation to climate change, organizing this connection in an effective and legitimate way seems to be of vital importance (Pielke, 2010).

The need for applicable knowledge in the governance of adaptation

Action research as co-production of science and policy

Current action research approaches have many different roots, and many sources have inspired the development and application of its methods (Reason and Bradbury, 2001). Action research starts from the idea that scientific knowledge has to be produced by creating, revisiting, and intervening in concrete social practices. In action research, the researcher generally enters a real-world situation and aims to both improve it and acquire knowledge (Checkland and Holwell, 1998). The aim is to build theories within the context of practice, and test them through some form of intervention (cf. Argyris and Schön, 1991). Action research aims both to contribute to the practical concerns of people in the field and to further the goals of social science simultaneously (cf. Gilmore *et al.*, 1986).

For now, it is enough to stress that action research essentially is a matter of co-production between practitioners and scientists: scientific knowledge is developed by designing, implementing, evaluating, and refining concrete interventions in concrete practices in close collaboration with these practices.

The ambitions of action research therefore fit nicely with the plea of many authors to enhance collaboration between scientists and policymakers in order to address the specific challenges of adaptation to climate change (e.g. Pielke, 2010; Pahl-Wostl, 2009; Hoppe, 2010). Pielke (2010) even claims that society's ultimate success in responding to, and preparing for, climate change in the face of ongoing uncertainty depends on a renewed relation between climate scientists and policymakers, based on the principles of co-production. Many others have pleaded for innovative knowledge arrangements that enable joint fact-finding or joint knowledge production (Ehrmann and Stinson, 1999; Edelenbos *et al.*, 2011). Chapter 2 presents a typology not only of different types of action research, but also of the various levels of interaction between researcher and research object that can be chosen to do action research.

The scientific promise of action-oriented research

The promise of action-oriented research is that the involvement of practitioners will enhance the development of actionable knowledge, and that researchers will provide the scientific underpinning of actionable knowledge and guard the development of scientifically sound theoretical knowledge. By engaging in complex governance systems, researchers are better able to understand their dynamics, increasing the research quality in terms of its sensitivity to contextual factors, the incorporation of local knowledge, and relevance. As Reason and Bradbury (2001: 9) stated, action research:

> lead[s] to 'better' research because the practical and theoretical outcomes of the research process are grounded in the perspective and interests of those immediately concerned, and not filtered through an outside researcher's preconceptions and interests.

By proposing alternative actions or strategies, researchers are able to ascertain the factors that explain behaviour, the barriers that people experience, and the belief systems that they hold. Therefore, action research is, from a scientific point of view, a promising approach because it results in a deeper understanding of practice. In this book, we critically reflect upon this scientific promise: does action research really result in a more profound understanding of what is happening in governance processes around climate change adaptation?

The normative starting point of action-oriented research

In general, action research cannot be neutral. By doing action research, the researcher tries to influence his or her object of research, not only to enhance insights, but also to improve the functioning of this object. In governance processes, action research can, for example, be aimed at improving the quality of stakeholder participation, the progress of planning processes, the extent to which policymakers are able to reflect upon their choices and their consequences, or at the smooth implementation of drafted adaptation strategies. In this book, we take it as a defining characteristic of good action research that the normative aspects are made explicit. The reader has to be able ascertain the researcher's normative position. In general, three normative ambitions regarding action research can be witnessed in this book:

- action research has to enhance policymakers' reflexivity: it has to enable them to reflect upon their own choices and behaviour by providing critical reflection on, or insight into, alternative possibilities;
- action research has to enhance the governance capacity necessary to formulate and implement adaptation strategies; this capacity exists partly in the competencies of involved people, but also partly in the institutional capacity of the arrangements at hand; and
- action research has to contribute to the legitimacy of climate change adaptation by improving provisions for public participation, science–policy interfaces, and collaboration between different stakeholders.

The various case studies reflect different normative starting points, and throughout the book we critically reflect upon these normative ideals and how they end up in practice.

Dilemmas of action research

In a context in which policy formulation is just starting and much is unknown about what constitutes effective and legitimate approaches, arrangements, and strategies, doing scientific research that is also policy relevant requires methods other than traditional (evaluative) case studies (in which practitioners are not actively involved and there is no aim to alter the situation). Action-oriented research is relevant not only for assisting policymakers by exploring what works, but also for analysing emerging policy processes that are meaningful to policymakers. In Chapter 2, the potencies of applying action research in the domain of climate change adaptation are described in more detail.

Although action research seems a very promising research methodology in the context of governance of adaptation to climate change, like with all research methods, several problems and pitfalls can be encountered in its application. Many of these pitfalls are also applicable when action research is applied in other policy domains, but some of them seem to be specific to the climate change adaptation domain.

First of all, there are fundamental differences between the research and policy institutions that have to be bridged in action research projects. Table 1.2 compares the logic of research and policymaking institutions on five aspects:

1 What constitutes progress in processes of conducting science and governance?
2 How is the scope for conducting science and governance defined and maintained?
3 Which influences contribute to adjustment and evolution within processes of science and governance?
4 What defines the type of interventions made to influence the course of events?
5 What outcomes are seen as valuable and effective?

These differences in institutional logic might cause problems in action research projects, and thus have to be dealt with. Also, the specific characteristics of, and assumptions underlying, action research might cause friction with the general assumptions of both institutions. In the various chapters, we further analyse the frictions and tensions between these logics. We expect to find at least three main tensions.

The processes of research and decision making are fuelled by different incentives which can be conflictive and mutually exclusive. Scientists are often confronted with internal pressure to publish in high-impact journals, because universities are increasingly working with performance indicators in which the number of publications in high-impact journals is a crucial element. This implies that scholars need to devote much of their time to writing and rewriting scientific articles. This is difficult to reconcile with collaborating with actors in the field. Also, high-impact journals tend to put strict demands on the rigour of the research, including a rigorous research design. Several journals are hesitant in accepting articles based on action research.

Table 1.2 Differences between the logics of research and the logics of policymaking (Termeer *et al.*, 2012)

	Research institutions	Policymaking institutions
Logic of progress	Empirical cycle: from research questions and hypotheses, to data collection, analysis, intervention, to evaluation	Disjointed incrementalism: non-linearity, hiccups, setbacks characterize decision process
Logic of structure	Research is organized around a single researcher and by establishing a structured set of involved participants	Decision-making processes are multi-actor, multi-level and multi-arena: elements of the process are located in different arenas
Logic of change	New data and insights are used to refine hypotheses and to adjust interventions	Changing circumstances or power balances can necessitate changing course
Logic of intervention	Interventions have to contribute to getting more insight into the way in which processes unfold and are aimed at testing theoretical hypotheses	Interventions have to contribute to realizing effective and legitimate collective action
Logic of outcomes	Results have to be scientifically valid and worth publishing	Results have to be politically feasible and have to attract enough resources to be implemented

Related to this tension between incentives is the issue of conflicting values. These may be ethical values, but also professional considerations. For a policymaker, it is important to enable political compromises, whereas scientists are often focused upon delivering the most effective solution. Policymakers are often confronted with time or budget constraints, whereas scientists would like to test hypotheses that require the mobilization of additional resources. These value differences make it difficult to define a common interest and goal.

Finally, research is essentially a goal-searching, exploratory activity, whereas policymaking is frequently organized in terms of narrowly defined projects in which existing insights are exploited, refined, and re-used. Also, the processes are structured differently, and this adds further to the difficulty of synchronizing processes of knowledge production and policy formulation.

Notwithstanding these problems, there have been numerous cases in which action research has been used with good results. Given the very positive reasons for using action research (more detailed insight into complex governance processes, increased social importance of research, and so forth), there is a lot to gain.

Aim and outlook

This book thus centres on action research for the governance of climate change adaptation and presents a variety of action research practices in this field. Chapter 2 (by Patrick Huntjens, Jasper Eshuis, Catrien Termeer, and Arwin van Buuren) introduces action research in more detail, explaining the various 'degrees' of action research. It forms the theoretical foundation for the other chapters. The main part of the book is composed of eight chapters that describe different studies in the field of governance of adaptation to climate change in which action research has played an important role. They describe the methodology of action research adopted, the problems encountered during the research process, the results of the methods (in terms of research results, knowledge utilization, and satisfaction of both practitioners and scientists), and a more general reflection upon the pros and cons of action research in the domain of governance of adaption to climate change. To get a good overview of different projects, two case studies are derived from the governance of adaptation to climate change consortium (GACC), three from other Dutch research projects, and five from projects in other countries, such as Denmark, Australia, and Vietnam. The topics studied in the case studies differ too, from flood management, land-use planning, and water management to the process of developing and implementing strategies.

Chapter 3 (by Martinus Vink, Daan Boezeman, Art Dewulf, and Catrien Termeer) couples Wittgenstein's ideas (Gasking and Jackson, 1967) on learning through the authentic view of a 'bad city guide' with the role that action research can play in puzzling over ideas and powering for support in the governance of climate adaptation. The authors describe an action research project in which they collaborated with a civil servant acting as guide in the policy network of the Dutch Delta Programme. Teaming up with Wittgenstein's 'bad' guide gave them insight into the array of actors' frames at the informal fringes of the network, yielding a complete picture of the wicked character of climate adaptation as a policy issue. They conclude that, for effective action research in policy networks, partnering with a guide is crucial not only for effective puzzling *over* the various practitioners' frames creating the problem, but even more so in terms of effective powering *with* practitioners' frames to gain a powerful say in the collective puzzle.

To develop a climate adaptation strategy for the Lower Vam Co River Basin in Long An Province, Vietnam, the VamcoPart Partners for Water project chose a participatory approach, based on the action research methodology. Given the Vietnamese culture and context, this could not be done in the same way as in Western cultures. The pilot project shows that action research methods – such as group model building (GMB) and highly interactive forms of learning – are possible in the Vietnamese context if properly embedded, initiated, and facilitated. Eventually, the project was able to find a way to connect with the Vietnamese participation tradition; and, in six meetings, more than 200 representatives of organizations at province, district, and commune level contributed to a series of GMB sessions focusing on a common understanding of problems, causes, solutions, and the development of strategy components.

This participative planning approach, together with advanced decision support tools, resulted in a Preferred Climate Change Adaptation Strategy. Chapter 4 (by Patrick Huntjens, Bouke Ottow, and Ralph Lasage) addresses the question of the extent to which action research methods can be applied in a non-Western culture like Vietnam, taking into account cultural differences and possible ways to bridge these.

In Chapter 5, Daan Boezeman, Martinus Vink, and Pieter Leroy elaborate how action research can be a particularly helpful way to make institutionalized ways of knowing, problem-solving, and decision making perceivable for a researcher. Institutional perspectives challenge purely rationalist approaches in stressing that actors interpret events in structures with which they are socialized. A strongly institutionalized context, where role expectations are stubborn, the science–policy interface strongly codified, and (potentially conflicting) competencies formalized, yields an interesting avenue to explore the potential for action research to deliver better grounded insights and societal changes. Their case study concerns the Dry Feet 2050 project, dealing with the future of the regional water system in the northeastern part of the Netherlands. Dry Feet 2050 aimed to organize knowledge production for climate adaptation in a more participatory way and engaged researchers to develop a joint action research project to enable learning thereon. Boezeman *et al.* used a number of action research methods: observing project meetings, organizing workshops on participatory governance and knowledge production, and reflection sessions with project members.

Chapter 6 (Rob Roggema, John Martin, and Lisa Vos) explains how design charrettes were used as a creative tool in participatory action research. The State of Victoria, Australia, wanted to develop knowledge on how to involve communities in decision making for climate adaptive futures and supported the research project entitled: Design-Led Decision Support for Regional Climate Adaptation. The design charrette methodology entailed an intensive multi-day and multi-participant design workshop aimed at creating innovative, creative, and integrated visions. The charrettes functioned as participatory action research in the complex arena of local/regional governance and climate adaptation. The researcher(s) interacted with a wide range of experts, local stakeholders, citizens, and policymakers. The chapter elaborates on the methodological advantages of this specific participatory action research, the results in terms both of climate adaptation visions and of participants' commitment and involvement, and evaluates and reflects upon the advantages and the disadvantages of the approach undertaken.

In Chapter 7, Gerald Jan Ellen, Corniel van Leeuwen, Wiebren Kuindersma, Bas Breman, and Frank van Lamoen study the difficulties that arise when adaptation strategies need to be implemented. The aim of the Adaptive Implementation Arrangements project was to develop – in interaction with stakeholders, universities, and knowledge institutes – a methodology to organize combinations of reflexive monitoring and flexible (legal, financial, and organizational) arrangements. The action research applied consisted of interviews and three types of meetings with stakeholders. The synchronization of science and practice was therefore sometimes difficult. However, the action research process led to a significant increase in reflection and learning between practice and science, also resulting in a higher degree of knowledge utilization and increasing the approachability of scientists for practitioners, and vice versa.

Patrick Driscoll and Martin Lehmann describe in Chapter 8 how the city of Copenhagen (Denmark) investigates new ways to develop, test, evaluate, and refine new forms of local governance tools, such as serious gaming. The city of Copenhagen uses these new tools for the implementation of their adaptation strategy. Driscoll uses a variety of action research tools to study this project, among which are interviews, recordings, focus groups, on-site observations, and discussions with project members.

In Chapter 9 Todd Schenk and Lawrence Susskind introduce role-play simulation exercises (RPS) as a powerful tool for supporting action research efforts. Their experiences with this type of serious game suggest that they can be invaluable when various stakeholders are engaged to collaboratively learn about climate change risks, explore options, and seek agreement on how to proceed with adaptive measures. Exercises constitute action research when officials and other stakeholders are actively engaged at all stages, from design to the interpretation of results, and the focus is on meeting community needs. Researchers working with communities can concurrently devise and test wider theoretical insights based on what happens during exercises and how participants reflect on their experiences. RPS exercises can be used to engage stakeholders in fictional yet realistic decision making that mimics challenges they are facing, or may soon face, allowing them to experience various dynamics and explore options in a low-cost, low-risk setting. Participants get a sense not just of the technical challenges posed by climate change, but also of the governance dynamics that make decision making difficult. This chapter draws on the authors' experiences of using RPSs around the world, and in particular with coastal communities in New England (United States) and infrastructure planners and decision makers in Singapore and Rotterdam.

In Chapter 10 Arwin van Buuren, Mike Duijn, Ellen Tromp, and Peter van Veelen describe a co-creation process that aimed to refine adaptive flood measures in such a way that they could be implemented. They describe how the process was executed by an interdisciplinary team of researchers and a policymaker from the City of Rotterdam. The chapter shows why action research in this case was a very useful approach and how the process was managed in order to deliver both policy-relevant knowledge and scientifically valid insights.

Chapter 11 by Mathijs van Vliet, Arwin van Buuren, and Jasper Eshuis includes an overall reflection on the use of action research in studying governance of climate change adaptation. It compares the difficulties hypothesized in this introductory chapter on the basis of the research presented in the eight cases presented in Chapters 3 to 10.

Acknowledgments

We want to thank all the authors of this book who participated in the Wageningen workshop for their comments on this chapter. Financial support from the Knowledge for Climate project, the Governance of Adaptation to Climate Change, and Erasmus University Rotterdam is gratefully acknowledged.

Note

1   A larger overview of adaptation projects in Europe can be found at http://climate-adapt.eea.europa.eu/. WeAdapt provides an overview of projects around the world, as well as downscaled climate data (http://weadapt.org/).

References

Adger, W.N., S. Agrawala, M.M.Q. Mirza, C. Conde, K.L. O'Brien, J. Pulhin, R. Pulwary, B. Smit and K. Takahashi (2007) Climate change 2007: impacts, adaptation and vulnerability. Contribution of working group II to the *Fourth Assessment Report of the Intergovernmental Panel on Climate Change*, Cambridge: Cambridge University Press, 719–743.

Argyris, C. and D. Schön (1991) Participatory action research and action science compared, in W.F. Whyte (ed.) *Participatory action research*, 85–96, Thousand Oaks: Sage Publications.

Arvai, J., G. Bridge, N. Dolsak, R. Franzese, T. Koontz, A. Luginbuhl, P. Robbins, K. Richards, K.S. Korfmacher and B. Sohngen (2006) Adaptive management of the global climate problem: bridging the gap between climate research and climate policy, *Climatic Change*, 78(1): 217–225.

Biesbroek, G.R., R.J. Swart, T.R. Carter, C. Cowan, T. Henrichs, H. Mela, M.D. Morcecroft and D. Rey (2010) Europe adapts to climate change: comparing national adaptation strategies, *Global Environmental Change: Human and Policy Dimensions*, 20(3): 440–450.

Boezeman, D., M. Vink and P. Leroy (2013) The Dutch Delta Committee as a boundary organisation, *Environmental Science & Policy*, 27: 162–171.

Brouwer, S., T. Rayner and D. Huitema (2013) Mainstreaming climate policy: the case of climate adaptation and the implementation of EU water policy, *Environment and Planning C*, 31(1): 134–153.

Bruin, K. de, R.B. Dellink, A. Ruijs, L. Bolwidt, M.W. van Buuren, J. Graveland, R.S. de Groot, P.J. Kuikman, S. Reinhard, R.P. Roetter, V.C. Tassone, A. Verhagen and E.C. van Ierland (2009) Adapting to climate change in The Netherlands: an inventory of climate adaptation options and ranking of alternatives, *Climatic Change*, 95(1–2): 23–45.

Buuren, M.W. van, P.P.J. Driessen, H.J.F.M. van Rijswick and G.R. Teisman (2014) Towards legitimate governance strategies for climate adaptation: Combining insights from legal, planning and democratic perspectives, *Regional Environmental Change*, 14(3): 1021–1033.

Checkland, P. and S. Holwell (1998) Action research: its nature and validity, *Systemic Practice and Action Research*, 11(1): 9–21.

Dewulf, A. (2013) Contrasting frames in policy debates on climate change adaptation, *Wiley Interdisciplinary Reviews: Climate Change*, 4(4): 321–330.

EC (2013) *Communication from the Commission to the European parliament, the council, the European economic and social committee and the committee of the regions: an EU strategy on adaptation to climate change*, Brussels: European Commission.

Edelenbos, J., M.W. van Buuren and N. van Schie (2011) Co-producing knowledge: joint knowledge production between experts, bureaucrats and stakeholders in Dutch water management projects, *Environmental Science & Policy*, 14(6): 675–684.

EEA (2013) Adaptation in Europe: addressing risks and opportunities from climate change in the context of socio-economic developments, EEA Report No 3/2013, Copenhagen: European Environmental Agency.

Ehrmann, J.R. and B.L. Stinson (1999) Joint fact-finding and the use of technical experts, in L. Susskind, S. McKearnan and J. Thomas-Larmer (eds) *The consensus building handbook*, 375–399, Thousand Oaks, CA: Sage.

Fankhauser, S., J.B. Smith and R.S. Tol (1999) Weathering climate change: some simple rules to guide adaptation decisions, *Ecological Economics*, 30(1): 67–78.

Gasking, D.A. and A.C. Jackson (1967) Wittgenstein as a teacher, in K.T. Fann (ed.) *Ludwig Wittgenstein: the man and his philosophy*, 49–55, New York: Dell.

Giddens, A. (2009) *The politics of climate change*, Cambridge: Policy Network.

Gilmore, T., J. Krantz and R. Ramirez (1986) Action based modes of inquiry and the host-researcher relationship, *Consultation: an International Journal*, 5(3): 160–176.

Hallegatte, S. (2009) Strategies to adapt to an uncertain climate change, *Global Environmental Change*, 19(2): 240–247.

Haug, C., T. Rayner, D. Huitema, R. Hildingsson, A. Jordan, E. Massey, S. Monni, J. Stripple and H. van Asselt (2010) Navigating the dilemmas of climate policy in Europe: evidence form policy evaluation studies, *Climatic Change*, 101(3–4): 427–445.

Hoppe, R. (2010) Lost in translation: a boundary work perspective on making climate change governable, in P.J. Driessen, P. Leroy and W. van Vierssen (eds) *From climate change to social change*, 109–130, Utrecht: International Books Utrecht.

Hulme, M. (2009) *Why we disagree about climate change: understanding controversy, inaction and opportunity*, Cambridge: Cambridge University Press.

IPCC (2012) *Managing the risks of extreme events and disasters to advance climate change adaptation: special report of the intergovernmental panel on climate change*, Cambridge: Cambridge University Press.

Jordan, A.J., D. Huitema, H. van Asselt, T. Rayner and F. Berkhout (eds) (2010) *Climate change policy in the European Union: confronting the dilemmas of mitigation and adaptation*, Cambridge: Cambridge University Press.

Leeuwen, C.W.G.J. van and M.W. van Buuren (2013) Connecting time spans in regional water governance: managing projects as stepping-stones to a climate proof delta region, in J. Edelenbos, N. Bressers and P. Scholten (eds) *Water governance as connective capacity*, 191-210, Aldershot: Ashgate,

Ministerie van Infrastructuur en Milieu (2013) *Klimaatagenda: weerbaar, welvarend en groen*, Den Haag: Ministerie van Infrastructuur en Milieu.

Pahl-Wostl, C. (2009) A conceptual framework for analysing adaptive capacity and multi-level learning processes in resource governance regimes, *Global Environmental Change*, 19(3): 354–365.

Pielke, Jr., R.A. (2010) Creating useful knowledge: the role of climate science policy, in P.J. Driessen, P. Leroy and W. van Vierssen (eds) *From climate change to social change*, 51–67, Utrecht: International Books Utrecht.

Pielke, R., G. Prins, S. Rayner and D. Sarewitz (2007) Climate change 2007: lifting the tabooon adaptation, *Nature*, 445(7128): 597–598.

Pralle, S.B. (2009) Agenda-setting and climate change, *Environmental Politics*, 18(5): 781–799.

Reason, P. and H. Bradbury (eds) (2001) *Handbook of action research: participative inquiry and practice*, Thousand Oaks, CA: Sage.

Termeer, C., A. Dewulf, H. van Rijswick, M.W. van Buuren, D. Huitema, S. Meijerink, T. Rayner and M. Wiering (2011) The regional governance of climate adaptation: a framework for developing legitimate, effective, and resilient governance arrangements, *Climate Law*, 2(2): 159–179.

Termeer, C.J.A.M., P.M.J.M. Huntjens, A.R.P.J. Dewulf, M.W. van Buuren and J. Eshuis (2012) Reconciling innovative knowledge partnerships into existing institutions, International Symposium 'The Governance of Adaptation', 22–23 March, Amsterdam.

Termeer, C.J.A.M., A. Dewulf, S. Karlsson-Vinkhuyzen, M. Vink and M. van Vliet (forthcoming) Changing governance and governing change: the wicked problem of adaptation to climate change, *Landscape and Urban Planning*.

Uittenbroek, C.J., L.B. Janssen-Jansen and H.A. Runhaar (2013) Mainstreaming climate adaptation into urban planning: overcoming barriers, seizing opportunities and evaluating the results in two Dutch case studies, *Regional Environmental Change*, 13(2): 399–411.

Verkerk, J., G.R. Teisman and M.W. van Buuren (forthcoming) Synchronising climate adaptation processes in a multilevel governance setting: exploring synchronisation of governance levels in the Dutch Delta, *Policy & Politics*. http://dx.doi.org/10.1332/030557312X655909

Verschuuren, J. (2013) Climate change adaptation under the United Nations Framework Convention on Climate Change and related documents, in J. Verschuuren, *Research handbook on climate change adaptation law*, 16–31, Cheltenham: Edward Elgar Publishing.

Vink, M.J., D. Boezeman, A. Dewulf and C.J.A.M. Termeer (2013) Changing climate, changing frames: Dutch water policy frame developments in the context of a rise and fall of attention to climate change, *Environmental Science and Policy*, 30: 90–101.

2 Forms and foundations of action research[1]

Patrick Huntjens, Jasper Eshuis, Catrien Termeer and Arwin van Buuren

Introduction

As described in Chapter 1, the core philosophy of our research approach can be described as developing a powerful combination between practice-driven research and theoretically informed scientific research. Practice-driven research means that we take guidance from the stakeholders in our case studies as the primary source of questions, dilemmas, and empirical data regarding the governance of adaptation, but also collaborate with these stakeholders in testing insights and strategies, and evaluating their usefulness. The purpose is to develop effective, legitimate, and resilient governance arrangements for climate adaptation. The ambition is to achieve scientific quality by placing this co-production of knowledge in a well-founded theoretical framework, and by involving partners working on climate adaptation in the field.

With action research becoming an accepted scientific methodology, many different approaches to it have blossomed. This chapter discusses the forms and foundations of action research, with the aim of clarifying its theoretical foundations. The next section gives an overview of the historical development of action research. The third section defines how we see action research in this book. In the fourth section, we distinguish its main current forms. The fifth section contains levels or intensities of action research, and the sixth section reflects on its scientific quality by dealing with the issue of recoverability. Then the chapter deals with ethical considerations and normative aspects with regard to action research. The chapter concludes by presenting the analytical framework used in the empirical chapters of this book.

Background, roots, and theoretical sources of action research

Action research has a rich history with several origins. It can be traced back to the social experiments that Kurt Lewin carried out in the 1940s (Lewin, 1946). Lewin's research on organizational change and social democracy explicitly aimed at social action. Other origins of action research can be seen in the Marxist idea that the main goal is not understanding the world but rather changing it (Reason and Bradbury, 2001). Paulo Freire's work on counter-hegemonic knowledge development together with oppressed people is one of the early forms of action research rooted in Marxist ideas (Freire, 1970). It has informed later participatory research aimed at emancipation and liberation of the underprivileged. Such research has been developed and implemented in, for example, participatory rural appraisal, educational research, and feminist research in different fields of practice (Reason and Bradbury, 2001). Another main source of action research is psychotherapy, where it has been used to develop forms of mutual inquiry and self-help. Within the fields of organizational change and leadership also, there is a history of action research. Under the flag of action research and action science, scholars such as Argyris (1985) and Torbert (1989) have built upon Lewin's work. More recent publications show that action research continues to be used in a wide range of disciplines and fields of research. Among others, scholars in organization studies (*engaged scholarship*, Van de Ven, 2007), social studies of science (Stirling, 2008; Wynne, 2006), and education studies (Stringer, 2004) are increasingly giving attention to how to engage problem holders in research projects.

In theoretical terms, action research draws on many sources. It builds on critical theory, humanism, feminism, constructionist theory, systems thinking, and complexity theory (cf. McIntyre, 2008; Reason and Bradbury, 2001). For example, critical theory informs action research in the sense that it aims at social change, and that it addresses power relationships influencing both practitioners and researchers in their practices and institutions (see e.g. Kemmis, 2001). Constructionist theory has added the idea that people learn most effectively by doing, and engaging in action. Constructionist theory stresses that learning is about constructing ideas by the one who learns, rather than teachers transmitting knowledge to pupils. Systems thinking is a grounding of action research when it comes to propagating holism and critiquing reductionist approaches (e.g. Checkland and Holwell, 1998; Flood, 2001). Systems thinking has revealed that solving problems in (complex) systems requires an understanding not only of the separate components of the system, but also of their interrelationships and their relation to the whole. Feminist theories have added to emancipatory goals of action research through their focus on making structures of domination visible and aiming to raise consciousness about those structures among men and women (McIntyre, 2008).

Influenced by the above-mentioned theoretical sources, action research has developed into a sophisticated research approach, applied in many disciplines with a rich variety of methods and tools. In the next section, we describe in more detail what action research exactly is, and how it can be distinguished from other research approaches.

What is action research?

Towards a definition of action research

As stated in Chapter 1, in action research the researcher enters a real-world situation with the aim of improving the situation and acquiring knowledge

(Checkland and Holwell, 1998). Although there are different strands of action research, such as action learning, action research, action inquiry, participatory action research, and collaborative action research (Eden and Huxham, 1996), all of them share the aim of building 'theories within the practice context itself and testing them through intervention experiments' (Argyris and Schön, 1989: 86).

A useful overall definition of action research is provided by Waterman *et al.* (2001: 4):

> Action research is a period of inquiry, which describes, interprets and explains social situations while executing a change of intervention aimed at improvement and involvement. It is problem-focused, context specific and future-orientated. Action research is a group activity with an explicit value basis and is founded on a partnership between action researchers and participants, all of whom are involved in the change process. The participatory process is educative and empowering, involving a dynamic approach in which problem-identification, planning, action and evaluation are interlinked. Knowledge may be advanced through reflection and research, and qualitative and quantitative research methods may be employed to collect data. Different types of knowledge may be produced by action research, including practical and propositional. Theory may be generated and refined and its general application explored through cycles of the action research process.

Action research aims both to contribute to the practical concerns of people in a problematic situation and to further the goals of social science simultaneously (Gilmore *et al.*, 1986). In other words, there is a dual commitment in action research to study a system and concurrently to collaborate with members of the system in changing it in what is together regarded as a desirable direction. The twofold ambition of developing practically relevant and scientifically sound knowledge requires the active collaboration of researcher and client, and thus it stresses the importance of co-learning as a primary aspect of the research process (Gilmore *et al.*, 1986). Action research involves utilizing a systematic cyclical method of planning, taking action, observing, evaluating (including self-evaluation), and critical reflecting prior to planning the next cycle (O'Brien, 2001). Therefore, both qualitative and quantitative methods can be used.

As clarified above, an important aim of action research is to develop actionable knowledge (Coghlan and Brannick, 2002). Several elements in action research help to provide knowledge that is relevant for policymakers. Firstly, the involvement of practitioners facilitates good access to the field and helps to gather rich data relatively easily, thus enhancing its usefulness (see for example Steins, 1999). Further, practitioners may help to formulate relevant research questions and demarcate the research object in such a way that it fits with their needs. Because data are gathered in context, the research results are bound to be valid in that context.

Table 2.1 Differences and similarities between action research, case studies, ethnography, and consultancy

	Action research	Case study	Ethnography	Consultancy
In situ research	Yes	Yes	Yes	Yes
Aim of social action	Yes	No	No	Yes
Researcher participates in action	Yes	Sometimes	Yes	Yes
Stakeholders participate in research	Yes	No	No	Sometimes
Scientific method	Yes	Yes	Yes	No

Differences from, and similarities with, other research approaches

We can further clarify what action research is by contrasting it with other research methodologies and with consultancy. As Table 2.1 shows, action research has several similarities with case studies (Blatter and Haverland, 2012; Yin, 1984) and ethnographic research (e.g. Burawoy *et al.*, 1991; Wacquant, 1995). It shares with those methodologies the element of the research being carried out in situ (in the midst of the action). It shares with ethnography and participant observation the element of the researcher participating in the activities and developments that are being studied. A main difference from both approaches is that action research aims to contribute to social action, but this is not necessarily a goal in case studies and ethnographies. These two aim at understanding and knowledge development, but they need not be aimed at actionable knowledge. Another difference is that in action research not only does the researcher participate in stakeholders' activities, but also stakeholders participate in research activities.

Differences from, and similarities with, consultancy

In practice, many types of consultancy projects use methods and tools for action research, but they do not necessarily use (or merit) the label of action research. The toolbox for action research (Huntjens *et al.*, 2011) includes quite a number of methods and tools that are also used in a non-action research context, albeit not embedded in an action research methodology, as practical tools for knowledge elicitation and/or process facilitation by consultants, policymakers, NGOs, and other practitioners. Hence, it is useful to identify some important differences and similarities between action research and consultancy.

One of the major similarities between an action researcher and a consultant is that both have an intense relationship or interaction with a problem holder (e.g. client or customer), compared to researchers who deliberately distance themselves from the problem holder. There are three possible relationships of an action researcher/consultant with the problem holder:

1   the action researcher/consultant examines the situation and provides the client (the problem holder) with solutions;
2   the action researcher/consultant helps the client (the problem holder) by jointly taking measures that are expected to have an effect; and
3   the action researcher/consultant intervenes independently to solve the problem for the client.

The main differences between consultancy and action research are:

1   consultancy does not have the aim of scientific knowledge development by testing scientific assumptions or by developing theoretically sound knowledge;
2   consultancy does not usually involve the use of a scientific research methodology that aims to ensure the recoverability and validity of the research; and
3   consultancy does not aim to have an effect in the scientific community.

Five approaches to action research

Within the family of action research, there are different orientations towards the main goal of action research (empowerment, transformation, social action in general), the role of those involved (from practitioners to co-researchers), the role of critique (focus on critique or on appreciation and positive development), and the degree to which the research is evaluative (from inquiry to evaluation). These different orientations can be traced back to five main approaches to action research: (a) cooperative inquiry, (b) participatory action research, (c) action inquiry, (d) appreciative inquiry, and (e) learning evaluation (cf. Edelenbos and Van Buuren, 2005; Ludema *et al.*, 2001; Reason, 2003). In the sections below, we draw extensively on the work of Reason and Bradbury (2001). We use the work of Edelenbos and Van Buuren (2005) to explain learning evaluations.

Cooperative inquiry

In cooperative inquiry, everybody involved in the research is a co-researcher and also a co-problem holder. As a co-researcher, everybody involved has a role in generating ideas, designing and managing the research, interpreting the results, and drawing conclusions (Reason, 1999). As co-problem holders, everybody engages in the activity under research (Reason, 1999). Cooperative inquiry can be applied as a form of democratic research with the explicit aim of cooperative inquiry to make research a democratic activity, giving both the practitioners and researchers a say in the research. As Reason (1999: 207) argues, it can be used to help 'ordinary people regain the capacity to create their own knowledge'. In that case, co-inquiry aims at emancipation. However, co-inquiry can also be used for more pragmatic purposes such as the enlargement of the research capacity or the enhancement of actors' learning by their being actively involved in the research process. The most important feature of cooperative inquiry is that the divisions between researcher and practitioners or between researcher and problem holder become blurred.

Participatory action research

Participatory action research (PAR) stresses political aspects of knowledge development (see e.g. McIntyre, 2008; Reason and Bradbury, 2001). It aims at conscientization[2] and enlightenment, but it also goes further in aiming at empowerment and liberation from oppression (Fals-Borda and Rahman, 1991). Researchers conducting action research in the PAR tradition explicitly choose sides. They do not aim to take a neutral or objective stance. One starting point of participatory action research is that it aims to improve the position of certain (disadvantaged) groups in relation to institutionalized power. In the field of climate change, participatory action research could, for example, aim at giving certain groups that tend to be overlooked or suppressed a say in climate change projects, for instance farmers, fishermen, or citizen groups. It often has an explicit ideological goal. A second characteristic of PAR is that it starts from the lived experiences of people (Reason, 2003). The (experiential) knowledge of the groups being researched is highly valued. This brings us to the third starting point of genuine collaboration, which is rooted in the traditions of the people involved. Thus the traditions, interests, and ideas of the research participants are to be respected and honoured.

Action science

Action science and action inquiry aim to develop effective action in the sense that they contribute to the transformation of organizations and communities (Reason, 2003). An important issue in action science is identifying 'the theories that actors use to guide their behavior' (Reason, 2003: 273). In the context of governing climate change, this could refer to, for example, the policy theories that actors use (theories about the relations between the problem, the means or policy instruments, and the outcomes). Therefore, the action researcher tries to discover both the *espoused theories* that actors claim to follow, and the *theories-in-use* that are actually being followed. The theories-in-use can be reconstructed by reflecting on action. Argyris and Schön (1978) have argued that such reflection can be aimed at action strategies (single-loop learning) but also at the mechanisms and variables that underlie action (double-loop learning). As is the case with other forms of action research, action science takes place in the midst of the action developed by the organizations and communities that are being studied.

Appreciative inquiry

Researchers engaging in appreciative enquiry start with *unconditional positive questions* in order to gain understanding of successes and best practices (Ludema *et al.*, 2001). Appreciative inquiry thus differs from critical approaches that are problem oriented and focus on deficits. Similar to other forms of action research, appreciative inquiry aims to contribute to social action. Different from other approaches in action research, it assumes that the most effective way of contributing to social action is to inquire into moments of exceptional enthusiasm, excellence, innovation, and beauty (Cooperrider and Srivasta, 1987; Ludema *et al.*, 2001). The idea is that positive elements are crucial to the vitality of organizations and networks, and, by researching and understanding those, one can effectively understand, sustain, and enhance such vitality (Cooperrider and Srivasta, 1987; Ludema *et al.*, 2001). Focusing on critique and problems is seen as a detour, which also runs the risk of being demotivating. Appreciative inquiry asks questions such as: What do you value most about your organization? What are best practices within your programme? (Ludema *et al.*, 2001).

Learning evaluation

Learning evaluations aim to improve policies and projects as they unfold during implementation (Edelenbos and van Buuren, 2005). Thus, learning evaluations are a continuing form of evaluation, differing from ex-ante or ex-post evaluations (cf. Scriven, 1991). In the context of governing climate adaptation, an advantage of continuing evaluation is that it is suitable for monitoring policies during implementation, thus providing information that can be directly used to adapt the ongoing policy process. Learning evaluations have a function of assessment, but also learning. Crucially, learning evaluation is a participative form of evaluation; users (the evaluated) and executors of evaluation (the evaluators) shape the evaluations in close interaction and consultation. An important element is the existence of frequent cycles of observation, conclusion, and (re) action. Observation and conclusion are not the end of an evaluation. A dominant element in the role of an evaluator is to be a reflective practitioner (Schön, 1983). The evaluator is closely involved in the policymaking process, and in a way is even part of it. The evaluator does not relate to his/her environment in an impersonal manner. In uncertain and unique situations, for which standard solutions are not available, he/she needs to contribute in a reflexive way to this policy context where he/she is part of the policy practice. The evaluator is in constant interaction with the actors he/she is evaluating. They must respond to the intermediate conclusions, after which the evaluator will determine their effects. Alkin (1990: 74) calls this *situated responsiveness*. This makes learning evaluation a type of action research. Action researchers are clearly oriented towards helping the policy practice they investigate and making a contribution to its improvement together with the actors involved (Stringer, 2004; Wadsworth, 2001).

Table 2.2 Main differences between five approaches to action research

Approach	Main goal	Key characteristic
Cooperative inquiry	Can be democratization of pragmatic	Division between researcher and practitioner becomes blurred
Participatory action research	Conscientization, enlightenment, and emancipation	Aims to improve the position of disadvantaged groups
Action science	Identifying the theories that actors use to guide their behaviour	Reflection on action strategies (single-loop learning) and mechanisms that underlie action (double-loop learning)
Appreciative inquiry	Contribute to social action through enthusiasm and stressing positive elements	Draws on positive developments (instead of critical reflection)
Learning evaluation	Evaluation and learning	Constant interaction between evaluator and evaluated

Choosing an approach to action research

The approaches summarized in Table 2.2 all have their merits, and it may not be easy to determine what approach to choose when one is considering action research. An important criterion is the goal that one is trying to realize through action research. Important questions are whether the main goal is emancipatory or not. If it is, PAR is a suitable option. If the main goal is evaluation, learning evaluation may be fitting. If one is aiming at reflection and reflective learning, action science, but also forms of PAR and learning evaluation, would be appropriate. Another important criterion pertains to stakeholders' preferences or capacities regarding their willingness to participate in action research, and their willingness to critically reflect on ongoing practice (this may be related to political sensitivity, but also to actors' institutional positions).

In practice, a mix of the approaches will usually be developed to fit the specific goals and preferences of the actors involved.

Levels of action research

Not only are there various approaches to action research, there are also different levels of intensity with regard to action research. This intensity has to do with two factors:

- the extent to which researchers and practitioners interact with one another, including the width and the depth of interaction (cf. Edelenbos and Klijn, 2006); and
- the extent to which researchers are actually involved in their object of empirical study.

With regard to the level of interaction, we distinguish two main dimensions: the breadth of interaction and the depth of interaction (cf. Edelenbos and Klijn, 2006). The breadth of interaction refers to the question of with whom the researcher interacts. The broader the interaction, the wider the selection of types of actors with whom the researcher interacts. Loosely based on Fung (2006), the following breadths of participation can be distinguished: interaction with only selected expert administrators and/or elected representatives, interaction with selected professional stakeholders of all kinds, interaction with selected professional stakeholders and lay stakeholders, interaction with self-selected stakeholders (open to all). For the depth of interaction, we distinguish four levels:

1 *Information*: researchers inform practitioners about their research plans and about their results.
2 *Consultation*: researchers consult practitioners about their main choices and about the validity of their results.
3 *Co-decision*: researchers and practitioners jointly decide about research questions, methods, and the way in which the results are formulated.
4 *Co-production*: researchers and practitioners work together in developing and executing the research process from beginning to end.

Although variation is possible in the field of action research, it is fair to say that the minimum level of interaction before we can speak about action research is consultation. In the case of researchers merely providing information, practitioners have no actual say in the research, and therefore this cannot be considered action research. In many cases, co-decision is necessary to realize real forms of collaboration and effective interaction that maximize joint learning.

Regarding the extent to which the researchers are involved in practice, we can distinguish between five levels, as set out in Table 2.3.

Genuine action research implies more than observation. However, there is huge variety when it comes to the other levels. There are many forms of collaborative investigation like brainstorming sessions, focus group meetings, and group model building. Learning evaluation can be seen as a form of collaborative action research on the level of reflection. Reframing is a clear example of intervention as level of involvement; and experimentation as a method reflects the most far-reaching level of involvement.

Determining an appropriate role for the action researcher

Directly linked to above considerations is the importance of an appropriate role for the action researcher. Upon invitation into a domain, the outside researcher's role is to implement the action research method in such a manner as to produce a mutually agreeable outcome for all participants, with the process being maintained by them afterwards. Accomplishing this may necessitate the adoption of many different roles at various stages of the process (adapted from

Table 2.3 Level of involvement during action research

Level	Action		Explanation
	Width of interaction	*Depth of interaction*	
Level 0 (not action research)	Selected co-researchers	Observation	There is no actual intervention but only (unobtrusive) observation of what is going on
Level 1	Selected expert administrators	Participatory observation	Researchers take part in the practices they observe, but they do not explicitly intervene in the situation to change practices and processes
Level 2	Selected professional stakeholders (incl. administrators)	Reflection	Based upon their observation and analysis researchers give their feedback to practitioners in order to improve practice
Level 3	Selected professional and lay stakeholders	Intervention	Researchers develop theory-based interventions in order to test hypotheses and assumptions
Level 4	Interaction open to all actors (self-selection by actors)	Experimentation	Researchers develop theory-based interventions in order to test hypotheses and assumptions

O'Brien, 2001), including those of planner, leader, catalyst, facilitator, teacher, designer, listener, observer, synthesizer, and reporter. Also, different roles can be divided within a researcher team. For example, one researcher may take up a role as facilitator of a change process, whereas another researcher from the same team may fulfil a more reflective or supervisory role. According to O'Brien (2001), the main role of an action researcher is to nurture local leaders to the point where they can take responsibility for the process. This point is reached when they understand the methods and are able to carry on when the initiating researcher leaves. In many action research situations, the hired researcher's role is primarily to take the time to facilitate dialogue and foster reflective analysis among the participants, provide them with periodic reports, and write a final report when the researcher's involvement has ended (O'Brien, 2001).

It is necessary to think about that dual role and to carefully negotiate entry into the situation and the researcher's role in relation to that of participants. Work to effect change and 'improvement' (as judged by people in the situation) can then ensue, with the researcher, however his or her role is defined, also committed to continuous reflection on the collaborative involvement and its outcomes (Checkland and Holwell, 1998).

Recoverability

However, action research remains an academic endeavour and thus has to correspond to academic standards. Traditional requirements for scientific knowledge development seem not to be applicable in a situation in which researchers strive for application-oriented knowledge. Natural science's strong card is repeatability, meaning that, in any scientific work (i.e. based on repeatable analysis published in a peer-reviewed journal), the research carried out needs to repeatable by interested outsiders. Because action research is often developed in complex situational contexts, where actors engage in active processes of interpretation and construction of reality (Ruggie, 1998), the research results are valid in that specific context. Moreover, during the process of action research, open dialogue may unlock untapped knowledge, generate new skills and know-how, produce higher quality reasoning for more legitimate policies, and create new, more collaborative interrelationships among the parties to the deliberation (Elster, 1998). This will make the repeatability of actors' behaviour unlikely, and action research less reliable than lab experiments. Nevertheless, action research may have a stronger truth claim than mere plausibility by making action research recoverable (vs. repeatable). Hence, action researchers need to be rigorous in their action research methodology, leading to scientifically sound research. Recoverability will help to justify the generalization and transferability of results from action (or case study) research. Recoverability is based on a declared-in-advance methodology (encompassing a particular framework of ideas) in such a way that the process is recoverable by anyone interested in subjecting the research to critical scrutiny (Checkland and Holwell, 1998). Hence, a seriously organized process of action research can be made to yield defensible generalizations. In summary, action researchers investigating social phenomena must at least achieve a situation in which their research process is recoverable by interested outsiders. In order to do this, it is essential to state the epistemology (the set of ideas and the process in which they are used methodologically) by means of which the researchers will make sense of their research, and so define what counts for them as acquired knowledge (cf. Checkland and Holwell, 1998).

Ethical considerations

Because action research is carried out in real-world circumstances, and involves close and open communication among the people involved, the researchers must pay close attention to ethical considerations in the conduct of their work. On the basis of work by Winter (1996), O'Brien (2001), Eversole (2003), Termeer and Kessener (2007), and Werkman *et al.* (2009), we draw attention to the following considerations.

Influence of stakeholders on the research

Action research aims to intervene in practice. This makes it even more important to give stakeholders a say in the research. Thus, it is important to consult the relevant stakeholders and take into account their preferences. The principles guiding the work should be accepted in advance by the stakeholders. As Winter (1996: 16–17) puts it 'All participants must be allowed to influence the work, and the wishes of those who do not wish to participate must be respected'. O'Brien (2001) argues that decisions made about the direction of the research should be collective. To this we would like to add that, in a governance context, it may not always be possible to gain consensus regarding every step in the research. More important than realizing consensus in every step is that the parties involved agree on the way of deciding on important issues (who should be involved, should the decision be taken by consensus or by majority, and so forth).

Transparency

Interlinked with the issue of giving stakeholders a say in the research is the idea that stakeholders should be able to follow and monitor the ongoing research. Thus, an important consideration is that the development of the research should remain transparent to the stakeholders. It may require extra efforts from the actors involved to ascertain that all actors actually have access to the information generated by the process, for example in the case of actors who have no direct access to scientific libraries or particular internet sources. In addition, researchers should be clear and open about the nature and aim of the research process, including personal preferences and interests (O'Brien, 2001).

Ownership of the research and the research products

Ethically, it is relevant not only that researchers should gain permission and consult stakeholders about decisions directly pertaining to the ongoing research, but also that stakeholders should be asked for permission if researchers want to collect or use data for purposes other than the specific action research on which they are working. Also, descriptions of others' work and points of view must be negotiated with those concerned before being published (O'Brien, 2001).

Confidentiality

The researcher has responsibility for maintaining confidentiality (O'Brien, 2001). This means that, unless the problem holders explicitly agree otherwise, it should not be possible to discover their identities on the basis of research reports or other research outputs.

Room for reflection

Action research implies that the researcher engages in the processes that he/she is studying and that the researcher is committed to, and involved in, action that adds to problem solving in practice. Although the researcher must be committed to facilitating change and dealing with practical problems, it is important that the researcher plays a role that is different from the role of practitioners, otherwise the added value of the researcher becomes less. Researchers may be of value at times when they bring in new ideas and they are able to reflect on the ongoing processes. One condition that facilitates such reflection and feedback by researchers is the opportunity to distance themselves physically and mentally from the ongoing processes on a regular basis, for example by regularly leaving the field and returning regularly to their university campuses.

Framework for analysis

On the basis of the above, we conclude this chapter with the framework used to position the action research in the following chapters.

Table 2.4 shows how various the goals of action research can be, and how diverse the interaction between the researcher and practitioners can be. The empirical chapters explore how these variations work out to effect the development of scientific findings and the practice of the governance of climate adaption.

Table 2.4 Framework for analysis

Main goal	Depth of interaction between researchers and practitioners	Width of interaction between researchers and practitioners	Level of researchers' involvement
Theory development – action	Information	Selected co-researchers	Observation
Inquiry – evaluation	Consultation	Selected expert administrators	Participatory observation
Reflection – emancipation	Co-decision	Selected professional stakeholders	Reflection
Prescription – intervention development	Co-production	Selected professional and lay stakeholders	Intervention
Prescription – theory testing	Co-production	Interaction open to all actors (self-selection by actors)	Experimentation

Note

1 This chapter is based on a research report by Huntjens, Termeer, Eshuis, and van Buuren (2011) titled Position paper on collaborative action research: foundations, conditions and pitfalls. This research report was developed within the research programme, Knowledge for Climate.

2 Conscientization is a social concept, grounded in Marxist critical theory, that focuses on achieving an in-depth understanding of the world, allowing for the perception and exposure of perceived social and political contradictions

References

Alkin, M.C. (1990) *Debates on evaluation*, London: Sage.

Argyris, C. (1985) *Action science: concepts, methods, and skills for research and intervention*, San Francisco, CA: Jossey-Bass.

Argyris, C. and D. Schön (1978) *Organizational learning: a theory of action perspective*, Reading: Addison-Wesley.

Argyris, C. and D. Schön (1989) Participatory action research and action science compared: a commentary, *American Behavioral Scientist*, 32(5): 612–623.

Blatter, Y. and M. Haverland (2012) *Designing case studies: explanatory approaches in small-N research*, Basingstoke: Palgrave Macmillan.

Burawoy, M., A. Burton, A.A. Ferguson, K.J. Fox, J. Gamson, N. Gartrell, L. Hurst, C. Kurzman, L. Salzinger, J. Schiffman and S. Ui (1991) *Ethnography unbound: power and resistance in the modern metropolis*, Berkeley, CA: University of California Press.

Checkland, P. and S. Holwell (1998) Action research: its nature and validity, *Systemic Practice and Action Research*, 11(1): 9–21.

Coghlan, D. and T. Brannick (2002) *Doing action research in your own organization*, London: Sage.

Cooperrider, D.L. and S. Srivastva (1987) Appreciative inquiry in organizational life, in R. Woodman and W. Pasmore (eds) *Research in organizational change and development*, 129–169, Greenwich, CT: JAI Press.

Edelenbos, J. and M.W. van Buuren (2005) The learning evaluation: a theoretical and empirical exploration, *Evaluation Review*, 29(6): 591–612.

Edelenbos, J. and E.H. Klijn (2006) Managing stakeholder involvement in decision-making: a comparative analysis of six interactive processes in The Netherlands, *Journal of Public Administration Research and Theory*, 16(3): 417–446.

Eden, C. and C. Huxham (1996) Action research for the study of organizations, in S.R. Clegg, C. Hardy and W.R. Nord (eds), *Handbook of organization studies*, 526–542, London: Sage.

Elster, J. (ed.) (1998) *Deliberative democracy*, Cambridge: Cambridge University Press.

Eversole, R. (2003) Managing the pitfalls of participatory development: some insight from Australia, *World Development*, 31(5): 781–795.

Fals-Borda, O. and M.A. Rahman (eds) (1991) *Action and knowledge: breaking the monopoly with participatory action research*, London: Intermediate Technology Publications.

Flood, R.L. (2001) The relationship of 'systems thinking' to action research, in P. Reason and H. Bradbury (eds) *Handbook of action research: participative inquiry and practice*, 117–128, London: Sage.

Freire, P. (1970) *Pedagogy of the oppressed*, New York: Herder and Herder.

Fung, A. (2006) Varieties of participation in complex governance, *Public Administration Review*, 66(suppl. 1): 66–75.

Gilmore, T., J. Krantz and R. Ramirez (1986) Action based modes of inquiry and the host-researcher relationship, *Consultation*, 5(3): 160–176.

Huntjens, P., C. Termeer, J. Eshuis and M.W. van Buuren (2011) *Collaborative action research for the governance of climate adaptation: foundations, conditions and pitfalls*, KvK Key Deliverable 1a, theme 7, Utrecht: Kennis voor Klimaat.

Kemmis, S. (2001) Exploring the relevance of critical theory for action research: emancipatory action research in the footsteps of Jürgen Habermans, in P. Reason and H. Bradbury (eds) *Handbook of action research: participative inquiry and practice*, 91–102, London: Sage.

Lewin, K. (1946) Action research and minority problems, *Journal of Social Issues*, 2(4): 34–46.

Ludema, J.D., D.L. Cooperrider and F.J. Barrett (2001) Appreciative inquiry: the power of the unconditional positive question, in P. Reason and H. Bradbury (eds) *The handbook of action research*, 155–165, London: Sage.

McIntyre, A. (2008) *Participatory action research*, London: Sage.

O'Brien, R. (2001) An overview of the methodological approach of action research, in Roberto Richardson (ed.) *Theory and practice of action research*, 1–18, João Pessoa, Brazil: Universidade Federal da Paraíba.

Reason, P. (1999) Integrating action and reflection through co-operative inquiry, *Management Learning*, 30(2): 207–226.

Reason, P. (2003) Three approaches to participative inquiry, in N.K. Denzin and Y.S. Lincoln (eds) *Strategies of qualitative inquiry*, 261–291, Thousand Oaks, CA: Sage.

Reason, P. and H. Bradbury (2001) *Handbook of action research: participative inquiry and practice*, London: Sage.

Ruggie, J.G. (1998) *Constructing the world polity: essays on international institutionalization*, London: Routledge.

Schön, D.A. (1983) *The reflective practitioner: how professionals think in action*, London: Temple Smith

Scriven, M. (1991) *Evaluation thesaurus*, Newbury Park, CA: Sage.

Steins, N.A. (1999) All hands on deck: an interactive perspective on complex common-pool resource management based on case studies in the coastal waters of the Isle of Wight (UK), Connemara (Ireland) and the Dutch Wadden Sea, PhD thesis, Wageningen University.

Stirling, A. (2008) 'Opening up' and 'closing down': power, participation, and pluralism in the social appraisal of technology, *Science, Technology & Human Values*, 33(2): 262–294.

Stringer, E.T. (2004) *Action research in education*, Upper Saddle River, NJ: Pearson/Merrill/Prentice Hall.

Termeer, C.J.A.M. and B. Kessener (2007) Revitalizing stagnated policy processes: using the configuration approach for research and interventions, *Journal of Applied Behavioral Science*, 43(2): 256–272.

Torbert, W. R. (1989) Leading organizational transformation, in W. Pasmore and R. Woodman (eds) *Research in organizational change and development*, 83–116, Greenwich, CT: JAI.

Van de Ven, A.H. (2007) *Engaged scholarship: a guide for organizational and social research*, Oxford: Oxford University Press.

Wacquant, L.J.D. (1995) The pugilistic point of view: how boxers think and feel about their trade, *Theory and Society*, 24(4): 489–535.

Wadsworth, Y. (2001) Becoming responsive: and some consequences for evaluation as dialogue across distance, *New Directions for Evaluation*, 92: 45–58.

Waterman, H., D. Tillen, R. Dickson and K. de Koning (2001) Action research: a systematic review and assessment for guidance, *Health Technology Assessment*, 5(23): 1–166.

Werkman, R., C.J.A.M. Termeer and J.J. Boonstra (2009) Omgaan met 'gedoe', in G. Smid and E. Rouwette (eds) *Actieonderzoek: ruimte maken voor onderzoekende professionaliteit*, 76–93, Assen: Van Gorcum.

Winter, R. (1996) Some principles and procedures for the conduct of action research, in O. Zuber-Skerritt (ed.) *Directions in action research*, 9–22, London: Falmer Press.

Wynne, B. (2006) Public engagement as a means of restoring public trust in science: hitting the notes but missing the music? *Community Genetics*, 9(3): 211–220.

Yin, R. K. (1984) *Case study research: design and methods*. London: Sage.

3 Action research in governance landscapes

Partnering with city guides and gatekeepers

Martinus Vink, Daan Boezeman, Art Dewulf and Catrien Termeer

Introduction

For philosopher Ludwig Wittgenstein, problems were not made solvable by single theories. Instead, for in-depth understanding, one could at most be 'guided' through the various perspectives to a problem. As a guide in philosophical problems, he considered himself a rather 'bad' guide, with advantages however:

> In teaching you philosophy I'm like a guide showing you how to find your way round London. I have to take you through the city from north to south, from east to west, from Euston to the Embankment and from Piccadilly to the Marble Arch. After I have taken you many journeys through the city, in all sorts of directions, we shall have passed through any given street a number of times – each time traversing the street as part of a different journey. At the end of this you will know London; you will be able to find your way about like a Londoner. Of course, a good guide will take you through the more important streets more often than he takes you down side streets; a bad guide will do the opposite. In philosophy I'm a rather bad guide.
>
> (Gasking and Jackson, 1967: 51)

Wittgenstein's plea to understand problems through the authentic view of the problem holder touches upon the plurality of societal understandings that often construct ill-structured or wicked problems (Hisschemöller and Hoppe, 1995; Rittel and Webber, 1972). Therefore, the variety of societal understandings of climate change mean that societal adaptation to climate change is often referred to as a wicked problem (Dewulf, 2013; Hulme, 2009; Vink *et al.*, 2013a). To deal with these persistent and difficult-to-solve problems, the true problem is to *define* the problem in coherence with the plurality of understandings, thereby corresponding with Wittgenstein's plea to understand a problem through its side-street views rather than through a single main-street view.

Accordingly, this chapter evaluates the potential of partnering with a 'bad city guide' as an action research (AR) approach to dealing with the various side-street views that complicate societal adaptation to a changing climate. We ask the following research questions: 1) how can partnering with a city guide as AR method help to gain a better understanding of the landscapes of actors, views and positions that construct the wicked policy problem of adapting to climate change; 2) how can partnering with a city guide as AR method provide policy advice that fits the problem holders' authentic understandings of the problem and its surrounding governance landscape. We address these questions because we believe that AR is currently only marginally used in scholarly analysis of governance processes (Wagenaar, 2011), and we therefore feel the need to specify the added value of deriving more credible as well as more pluralistic understandings of the wicked problems and solutions often associated with the governance of climate adaptation.

Questioning AR in a governance context goes beyond the organizational context in which AR is usually employed. We therefore build on AR viewed from the systems thinking theory discussed in Chapter 2, which proposes that solving problems requires understanding the interrelatedness of actors in complex governance systems, in combination with AR viewed from constructionist theory to understand how the interrelatedness of actors may change through learning by interacting. We apply this system-learning approach by focusing on frames as short storylines that actors share in interaction to create a common understanding (Dewulf et al., 2009). In addition, we consider these frame interactions as the intermediaries between collective processes of puzzling over ideas and the collective processes of powering over interests (Heclo, 1974; Vink et al., 2013a; Vink et al., 2013b). To gain a better understanding of the potential of AR for scientific understanding and organization change in these processes of puzzling and powering, we studied frame reflections between policymakers and us as scientists (Checkland and Holwell, 1998; Coughlan and Coghlan, 2002). We did this in relation to the governance of adaptation to climate change (GACC) in the Dutch Delta Programme for Lake IJssel (in Dutch IJsselmeer) over the period 2010–2013. This chapter gives an account of how we participated in network meetings, paying special attention to our reflections with our guide in the network and the gatekeeper who gave us access to the network (Bache et al., 1996; Gasking and Jackson, 1967).

Next, we elaborate on our theoretical lens in doing AR on wicked policy problems through partnering with a guide. We elaborate on the nature of wicked problems, and the role that frame reflection might have in elucidating and altering processes of puzzling and powering over them. Subsequently, we sketch our case study and how we applied the theoretical lens in our methodological approach. After that, we discuss the results in terms of scientific understanding and organizational change, and we conclude that not only frame reflection is an important tool in AR, but partnering with Wittgenstein's bad city guide is also. We discuss the characteristics of this bad city guide in the scholarly tradition of boundary work (Boezeman et al., 2013; Clark et al., 2011; Gieryn, 1995; Jasanoff, 1994).

Action research: navigating wicked policy landscapes

As a quintessential long-term policy problem, the governance of adaptation to climate change relies on knowledge about long-term climate change impacts, which are riddled with uncertainties. In addition, this long-term character implies multiple policy cycles before impacts materialize and before the effects of adaptation measures can be evaluated. This can render policymaking over adaptation prone to controversies about the knowledge base and may amplify political conflict over long-term versus short-term interests (Hovi *et al.*, 2009; Lazarus, 2008; Lempert *et al.*, 2009). Some have referred to climate change adaptation as a wicked problem that cannot be precisely formulated or solved because of the uncertain knowledge in combination with the fact that diverging problem formulations change over time (Davoudi *et al.*, 2009; Jordan *et al.*, 2010; Lazarus, 2008; Rittel and Webber, 1972). Accordingly, the governance of adaptation to climate change might be characterized by (1) inherent uncertainties given the long-term character of this policy issue; (2) the association of many interdependent actors with their own ambitions, preferences, responsibilities, problem framings, and resources; and (3) the lack of a well-organized policy domain enhancing and monitoring long-term climate adaptation on the short-term policy agenda (Ford *et al.*, 2013; Termeer *et al.*, 2013)

Although the scientific attention on GACC has been steadily growing over the last decades, this has not led nation-states to unconditionally implement climate adaptation policies. Moreover, although a lot of the literature specifically stresses the need for state intervention, most of it adopts a rather abstract view on how rules and regulations should work in GACC, instead of working towards an in-depth empirical understanding of how GACC processes take shape. Hence, a growing body of scientific knowledge on the role of policy systems may not of itself lead to in-depth understanding of, and usable knowledge for managing, the complexity of GACC processes (Biesbroek *et al.*, 2010; Biesbroek *et al.*, 2013; Ford and Berrang-Ford, 2011; Keskitalo, 2010; Repetto, 2008; Vink *et al.*, 2013b).

Policymaking as an interplay of puzzling over routes and powering over destinations

In other domains, scholars have emphasized that to understand the empirical complexity of wicked policy problems neither a technocratic view of policymaking – where the 'best' policy option can be derived from proper calculation and modelling – nor a more political perspective on policymaking – where stakeholders negotiate over their interests on the basis of rational microeconomic thinking – are applicable. An interesting way of conceptualizing policymaking about wicked problems might be what Heclo (1974), Hall (1993), and Culpepper (2002) call a process of both 'puzzling and powering', where governance is about collective 'puzzling' over ideas and concepts to come up with plausible storylines and solutions, and at the same time about organizing enough 'power' to get things done in a plural societal context. Scholars like Stone (1989), Schön and Rein

(1994), Yanow (1996), and Wagenaar (2011) define this dichotomy between puzzling over ontology and powering over normativity by introducing the concept of framing to the policymaking process. They understand policymaking as a matter of negotiating over language, and with language, in which they consider frames short storylines or metaphors, explicitly or implicitly saying something about the cause of the problematic reality, and at the same time taking a moral standpoint towards this reality implicitly pointing towards possible solutions.

Framing climate adaptation, for example in the context of the low-lying Netherlands as *a national water task*, implies that climate adaptation is primarily a matter of water management at national level. The word *task* implies that something has to be done without a lot of room for debate. A frame also excludes possible side-street views on a problem. The possibility of mitigation as a solution to the problem is clearly omitted, as well as the possible variety in regional needs and opportunities for adaptation; the concept of *task* does not take into account plural understandings of how important adaptation is in the first place. In the social context of policymaking, like administrative meetings, parliamentary debates, or stakeholder workshops, frames like the *water task* might interact, compete, or merge with other frames. This process can be understood as the interactive alignment of societal understandings, or what Wittgenstein calls the various side-street views on a problem.

Action research: powering to puzzle

Following Heclo (1974), puzzling over climate change uncertainty and the collective wondering over adaptation options is clearly a process in which interactive framing plays an important role. To organize power to get climate adaptation policies implemented, interactive framing might also be important. Policy actors might strategically choose both the *partners* with whom they wish to interact, and the *frame* they employ in interacting with these partners to organize support. By framing issues in a strategic way towards (influential) stakeholders, administrators, experts or politicians, policy actors may strategically puzzle towards power. Framing climate adaptation as a national water task is more likely to fall on fertile ground at a ministry of public works responsible for executing nationally defined water tasks than in an international NGO concerned with reducing CO_2 emissions. This notion of governance as actors puzzling and powering closely resembles AR as a system-thinking approach in combination with a constructivist approach (Huntjens *et al.*, Chapter 2, this volume). For AR to be effective, it needs to focus on the holistic system of actors' puzzling and powering through frame interactions, and at the same time AR needs to interact in these processes for social learning to occur. If puzzling and powering are interplaying processes, this would mean that knowledge is not sovereign and that adding new perceptions to the policy puzzle by social learning might be met with scepticism because of its effect on policy coalitions and power constellations. Therefore, AR might have to organize power to get legitimate access to the governance process. This would mean that AR is first of all a matter of powering to puzzle.

In modern deliberative governance arrangements often proposed by the GACC literature (Vink *et al.*, 2013b), processes of puzzling and powering are less dependent on a sovereign regulating authority and more on a market-like co-production of equal players in a network. Nevertheless, government often plays an important role in guiding network governance as a gatekeeper, determining who is included and who is excluded (Bache *et al.*, 1996). This regulation is often ad hoc and not formalized (Goodwin and Grix, 2011; Rhodes, 1996; Stoker, 1998). Accordingly, AR in GACC may be not only about creating a shared understanding in a closed organizational or governmental context, but also, and primarily, about navigating and influencing processes of sense-making in loose networks of stakeholders (Coughlan and Coghlan, 2002). In this context, legitimate access to these networks is a first step, in which the networks' gatekeepers play an important role (Bache *et al.* 1996; Bache, 2000; Barzilai Nahon, 2008). To be effective in powering to puzzle, action researchers may be wise to start by creating shared understanding with a gatekeeper.

In addition, AR is about holistically understanding and navigating the landscape of actors, issues, and perspectives that shape the puzzling and powering processes in the network. When successful knowledge brokering depends on in-depth knowledge of network actors' positions and perspective, a network guide (Gasking and Jackson, 1967) might become essential for knowing where, when, and how to frame issues or with whom to interact to effectively broker a frame or side-street view on a problem. When policy networks represent wicked governance issues not all guides will do the same guidance and not all gatekeepers will do the same gatekeeping. Understanding and reflecting on both the guide and gatekeeper's framings of the network will become essential for good AR.

Action research methods applied in the Delta Programme IJsselmeer (DPIJ)

To illustrate our ideas on effective AR in wicked GACC, we elaborate on an AR project conducted in the GACC context of the Dutch Delta, a lowland region inhabited by 10–17 million citizens depending on the definition of the Delta, which has a long history of living with sea level rise (Warner *et al.*, in press). However, following increasing societal concern about climate change and sea-level rise in the early 2000s, a Dutch political advisory committee presented its advice on flood safety to the Dutch government in 2008 (Delta Commissie, 2008). This second Delta Commission adopted a rather wide perspective to its original task of reviewing flood safety. Together with some far-reaching recommendations on the Dutch institutional arrangements for flood protection, it set the agenda for a focus on fresh water availability during summer droughts, and pointed towards the country's largest freshwater lake and its potential capacity for buffering civic, agricultural, and industrial water demands (see Figure 3.1). Although the committee's recommendation resulted in little opposition in cabinet and parliament, at the geographical scale of the lake the proposal to raise the lake's water level by 1.5 metres to provide increased storage capacity touched

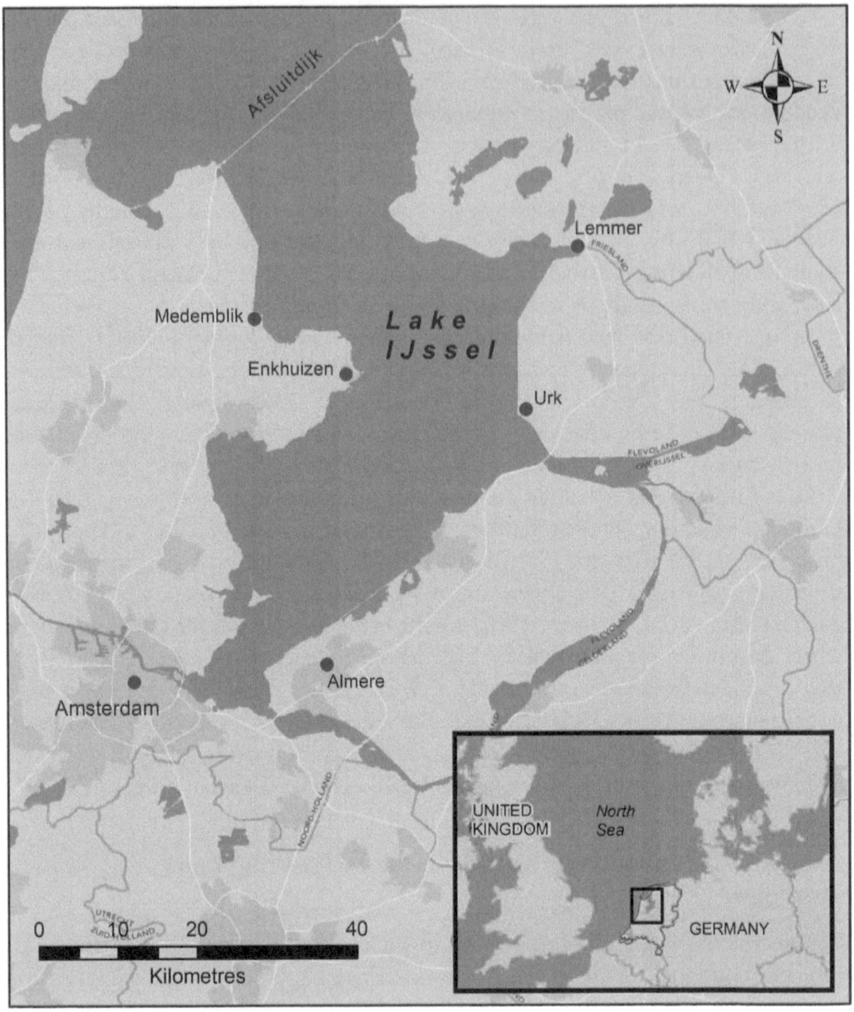

Figure 3.1 Map of Lake IJsel

upon a wide array of issues, ranging from potential inundation of industrial areas and picturesque waterfront towns to the failure of water management structures (Boezeman *et al.*, 2013; Vink *et al.*, 2013a). Accordingly, at lake scale, climate adaptation plans did not go unopposed.

For that reason, the Delta Programme for IJsselmeer (DPIJ), established two years later, adopted a bottom-up deliberative governance approach to operationalizing the committee's agenda in climate adaptation strategies legitimate for local governments and stakeholders. After a successful year of building a regional governance network occupied with mutual learning over the various issues and joint fact-finding for mutual understanding in the lake area, the governance process entered troubled waters. The initial sense of urgency faded among specific groups of the more than 200 public and private stakeholders. Other groups tended to take action outside the governance network by developing plans and taking positions on their own. DPIJ became a landscape of agencies, regulation, and actors around the wicked issue of rising water levels consequent to climate change. The landscape was connected through a newly established governance network, but still all actors maintained their official roles in their unique locations at different institutional settings and governmental scales. At this point, the director of the administrative office occupied with gatekeeping (Bache *et al.*, 1996) the governance process of DPIJ, approached us as researchers in the Knowledge for Climate Research programme, asking us to make sense of the governance process and to advise on possible routes towards effective and legitimate climate adaptation strategies.

Methodological approach to action research in DPIJ

In line with our conceptual notions about a guide, gatekeeper, and frame interactions in puzzling and powering over climate adaptation, we viewed AR through the theoretical lens of systems thinking in combination with a constructivist approach to learning. In practice, this meant that AR was about the understanding of, and interacting with, 'knowledge in action' (Coughlan and Coghlan, 2002) which went beyond a detached description of an institutional setting or controlled environment, but did not go as far as co-production of research questions with policy actors or us co-producing official policy documents. We specifically partnered with an administrator from the administrative office acting as our guide, who we initially consulted in our research design, and who in a later stage of the process co-decided with whom to interact. Initially, we conducted cooperative data inquiry with our guide, but in the later stage this was extended to frame reflection on actors' frames and corresponding behaviour, often referred to as action science as discussed in Chapter 2. By conducting cooperative inquiry and action science, we aimed to select more in-depth and context-specific data, and at the same time yield analyses that fit the policy actors' understanding and could be employed in the puzzling and powering process to transform policy action (Gasking and Jackson, 1967; Riordan, 1995).

For analysis and reflection, we approached textual frames using a constructivist social linguistic approach (Phillips and Hardy, 2002; Wood and Kroger, 2000). In this approach, we focused on textual frames in terms of 1) the textual interactions taking place between agents in a network and the linguistics shaping the textual interactions between the agents, and 2) the textual frames as presentations of crystallized meaning in documents or statements produced by these agents. In this approach, we consider textual interactions as meaning-making devices often unconsciously employed by the agent (Dewulf *et al.*, 2009; Schön and Rein, 1994; Weick *et al.*, 2005), but also as tools which may be used strategically for including, excluding, emphasizing, or downplaying issues, in a broader process of puzzling and powering (Benford and Snow, 2000; Entman, 1993; Heclo, 1974).

Data collection and frame reflection

To collect frame interactions and reflect on frame interactions in DPIJ, we used four research methods on an increasing scale of interaction in the governance process: 1) the collection of textual programme documents, 2) participatory observation, 3) semi-structured interviews, and 4) frame reflection meetings. Initially, we took a more distant observing approach akin to what Yanow and others have called *abduction* (Magnani, 2001; Yanow and Schwartz-Shea, 2006), where we iteratively switched between theory building and data collection through reading, observation, and interpretation with the occasional help of our guide. After this stage, we actively partnered with our guide for the more interactive part of the AR project.

Textual analysis of policy documents

To understand how frame interactions in puzzling and powering over climate adaptation crystallized in policy documents, we collected the (intermediate) policy documents produced by the administrative office after meetings or in preparation for new meetings. We collected all documents emanating from network meetings that we attended or which were used as input for discussions in network meetings.

Participatory observation

The participatory observation consisted of observing, listening, and taking notes of frame interactions in discussions and presentations at, in total, three IJsselmeerdays[1], three decision-maker conferences[2], four preparation meetings, and 10 meetings with administrative office staff[3]. If the administrative office agreed, the discussions were also recorded on a digital voice recorder. Due to political sensitivity and the closed and informal character at which the meetings were aimed, the administrative office generally did not support the idea of recording these meetings. All meetings were organized by the DPIJ administrative office, with a consultancy firm assisting in taking care of the logistics and the organization process. The administrative office consisted of both national civil servants – mainly from the Ministry of Public Works – and civil servants from county level or water board employees seconded to the administrative office. During the research, the administrative office consisted of about 20 people depending on when secondment contracts started or ended. In line with the iterative and cyclical character of AR (Coughlan & Coghlan, 2002), some of the frame analyses featured in the governance practice itself by providing ad hoc frame analysis feedback to practitioners. By asking reflective or clarifying questions, we orchestrated actor-centred analyses of frames employed. If statements or discussions remained implicit or unclear due to jargon or presumed common understanding among participants, we asked for clarification. In a later stage of the research, we took a more active approach.

Semi-structured interviews

We conducted 21 semi-structured interviews with practitioners throughout the governance network. The interview questions centred on the events and discussions that took place in network meetings, and how actors made sense of these events and discussions. We selected the 21 interviewees on the basis of their institutional role (civil servant, decision-maker, expert, private sector representative, civil society representative) and their position in the network in terms of distance to the administrative office. Using our guide's suggestions of who to interview, we strove for a selection of interviewees representative of a wide variety of frames, or Wittgenstein's side-street views (Gasking and Jackson, 1967). All interviews except one were recorded by digital voice recorder with the interviewees' permission. The interviews took between 50 and 90 minutes and were conducted by one researcher in cooperation with the guide who had functioned as a liaison in contacting the interviewees. Because of the official and sometimes politically sensitive character of the DPIJ, interviewing high-level civil servants and decision-makers would probably not have succeeded had we not been accompanied by our guide. Although this cooperation could suggest a bias in the wording of the questions or hesitance on the part of the interviewees about giving honest answers, the researcher who did the interviewing took great care to explain that the interview was open and without consequences. During the interviews, we continuously monitored the quality of the interview material by walking the respondents through the sequences of network events and discussions, which avoided generalizations, or expression of plain opinions or good intentions, and we avoided asking closed questions (Wagenaar, 2011: 253–258). By confronting interviewees with other framings of issues, processes, or problems, we provoked reflection on their own frames or behaviour. In addition, this provided the opportunity to test our own analyses of the governance process.

Frame analyses methods

After collecting these first frames, frame interactions, reflections, and theory testing, we started analysing. In line with our constructivist approach to frame interactions as processes of puzzling and powering in GACC, we relied on frame analysis. A common denominator among different varieties of frame analysis is the assumption that 'if we wish to understand social events, we need to look directly at those events as these unfold, not at retrospective reports or second-hand data or other forms of "self-report"' (Wood and Kroger, 2000: 26). Accordingly, texts are studied as parts of the concrete interaction context where they occur. This fits the textual data gathered during participatory observation at the DPIJ meetings and during frame interactions in interviews. For analyses, Wood and Kroger (2000: 91–95) offer a series of general guidelines for doing textual analytic research, a number of which are worth mentioning here: (1) try to identify the meaning to and for the participants; (2) do not ignore the obvious but try to explain it; (3) concentrate on what the speaker is saying; (4) explore the consequences of slightly different versions of the text through thought experiments; (5) look carefully at how the text is structured; (6) be alert for multiple functions of discourse; (7) adopt a comparative stance, (8) question the taken-for-granted; and (9) pay attention to grammar (e.g. passive versus active formulations).

Organizing frame reflection with our city guide, gatekeeper, and other policy actors

Having observed, read, interviewed, and conducted analysis in cooperative inquiry with our guide, we actively participated in the governance practice in the form of action science through seven organized frame reflection sessions with the civil servants from the administrative office and some representatives of stakeholder groups. We centred the reflections on the frames in use and the related behaviour of policy actors discovered in the previous phases. The reflections were organized during two administrative preparation meetings, and five meetings with the network's gatekeeper. Presentations and subsequent reflections on our preliminary findings and theory building resulted in discussion with the participants that served as reflecting moments for the governance actors involved, served as theory testing, or yielded new data for analyses (Friedman and Rogers, 2009). Finally, in an informal ad hoc way, our guide accompanied us always during the interactions with the DPIJ governance network and yielded constant action science at co-decision level (Huntjens *et al.*, Chapter 2, this volume) in the form of frame reflection and theory testing during more than 40 lunch meetings, car rides to interviewees, or coffee breaks.

Action research results

Textual analyses of frame interactions

After the director of the DPIJ programme approached us as action researchers to reflect on the DPIJ network process, we started with observation of the process during the various network meetings. This yielded the preliminary frame analyses. Where the administrative office framed the issue at stake as a rather technical exercise: 'doing nothing isn't an option', frame interactions between the administrative staff and the network during IJsselmeerdays and IJsselmeer summits did not arrive at a consensus in terms of what was, or what ought to be, at stake (Van Buuren *et al.*, in press; Vink and Mulligen, 2013). In preliminary policy documents however, the administrative framing prevailed. What did appear was the different scale at which actors framed the problem at stake, and whether the problem was framed as technical, procedural, or societal. The technical problem frame as presented by the administrative staff was mainly countered by regional public and private stakeholders through questioning the implicit priorities and assumptions embedded in the administrative framing: 'do these "corner points" [policy options] imply extra safety norms?', or 'you claimed that ecology is an autonomous process, is that actually the case?' and 'what strikes me is that freshwater demand calculated for is so big in relation to the current freshwater use, how did you determine this future demand?'. The counter-frames suggested that the problem presented by the administrative office was not only about the different technical scenarios of working towards the water task, but also possibly about the preferred safety norm, assumptions about future societal preferences, or different issues at stake not yet taken into account. This would suggest a need for societal debate.

Later on during the research project, discussions among administrators, societal representatives, and politically elected regional decision-makers intensified (Van Buuren *et al.*, in press; Vink and Mulligen, 2013). The issue at state at the Lake IJssel summit was framed by the administrative office as knowledge centred: '

> What do you decision makers think of what we prepared and what does the Director General from the Ministry of Public Works [who is present today] need to take to The Hague[4] [in terms of knowledge prepared]

This was countered by the Director General (DG) as follows:

> …what do I take with me to The Hague? Yes, but we are talking about a collective programme here, the national government is also owner, there are five decisions to be made in conjunction … we do not make decisions that stand on their own.

The administrative office framed the issue as their ongoing work on a water task, after which the DG countered this framing by framing the issue as a matter of who is in charge. Apparently working on a water task is not all that matters to the participants. The way this mismatch between the administrative office and the DG was subsequently framed by a politically elected regional decision-maker shows the tension between the scales at which the issue is being framed:

> So is this a mismatch between the administrative office and The Hague [national government]? Who is determining over what then? … Maybe The Hague should take care that the other [regional delta] programmes adapt to DPIJ?

The frame interactions that we observed during the IJsselmeerdays and Lake IJssel summit revealed considerable tension in problem framing and scale framing. We therefore used the analyses of these interactions as input for the interviews with the 21 selected practitioners from the governance network.

Interviews with practitioners

At this stage, we took an active role in the governance practice by actively reflecting with practitioners on their framing during interviews. Our active role was guided by our guide. Two things appeared to be important during these frame reflections. First, the selection of the practitioners: because as outsiders our knowledge of the governance landscape was limited, without our guide we would probably have selected officials and readily accessible administrators or stakeholders. An officially appointed guide like an administrator tasked with public relations or the DPIJ director herself would, for the sake of portraying the official view, probably have selected similar interviewees. However, our guide – a normal administrator with over 20 years' experience in the region, tasked with organizing the governance process – knew all the complex relations and, hence, the various side-street views (Gasking and Jackson, 1967) that were important for understanding the wickedness or complexity of the policy landscape. Employing our guide to select interviewees yielded a much more in-depth understanding of the governance praxis.

The second thing that appeared important was that during the interviews our guide functioned as a liaison between us as scientific outsiders and the official decision-making part of the DPIJ governance network. As part of the administrative organization of the network, our guide was in a better position to confront the interviewees with the administrative framing of the issue, and this resulted in frame reflections on the spot between the administrative framing of our guide and the various network members' framing. We as researchers could reflect on both framings on the spot if clarification was needed or if frame differences appeared persistent.

This guided form of interactions with governance practitioners yielded a wide range of all the side-street views that made up the wicked policy landscape. What appeared was a map of all frames within the governance landscape, showing differences in the scale at which practitioners framed their issues of concern, as well as the nature of these issues. Issues were framed in national technical ways:

> In the latest Delta Decisions the Delta Commissioner formulated the 1.5 metre water level rise in a very careful way to distinguish between sense and nonsense (director national Delta Programme).

In local socio-economic ways:

> The municipality is picking the things out of DPIJ that are of concern to the municipality; the municipality does not want to suffer from this programme, we have a shipyard and we want to focus on building coasters, so we wish to expand our harbour area, and thus we are protecting our current waterfront" (municipality alderman).

In procedural ways:

> The [national] water task becomes clear now ... but what you see is that our project organization starts to hassle, which regularly creates internal conflict, because the standard project organization with the traditional preconditions,

time horizons, and budgets cannot take into account these new long term issues (municipality administrator)

Or in relational ways:

The DPIJ administrators think they are neutral in their search for knowledge, but they aren't (societal organization representative).

During the interviews, reference was often made to an increasing apathy towards the nationally and technically framed water task as presented by the administrative office, which until that time held all options open and resulted in abstract forms of consensus on technicalities as presented in meeting reports, camouflaging a lack of frame interactions on societal concerns (Van Buuren *et al.*, in press; Vink & Mulligen, 2013). During the interviews however, our guide and the other practitioners did interact in relation to frame differences due to the small closed and informal setting. This promoted frame reflections on the part not only of the interviewees, but also of our guide as a member of the administrative office. Our guide's frame reflection appeared to be extremely valuable in organizing support for our findings at the administrative office at a later stage.

Frame reflections

Both the frame analyses of the various DPIJ network meetings and the frame reflections during the 21 interviews yielded an analysis of frame interactions as a constant interplay of puzzling and powering over different framings of the issue at stake in DPIJ. One of the main conclusions that we arrived at in cooperation with our guide was that the DPIJ's administrative office largely controlled the network discussions by setting the agenda in a nationally and technically framed form. Through de-politicization, discussions became knowledge centred and procedural, and this fitted well with the task of the administrative office. In other words, the administrative office had empowered itself into a central role in the puzzling process by framing the problem as technical or procedural, omitting any more political discussion or negotiation over socio-economic or political frames that might have been conflictive at different scales. On the other hand, this resulted in various practitioners struggling with how to frame their local societal or economic concerns to fit the national technical frame. In the end, this led to the previously mentioned increasing apathy.

Analysing a governance landscape through the eyes of a guide showing the various side-street views that determined the wicked character of the landscape was one thing. Powering our analyses into the administrative office's puzzling process was a second thing. Because the director of the office, who acted as gatekeeper to the network, tasked us to do the analyses and at the same time had a central role in setting the agenda, we reported back to her. To disseminate the analyses and provoke reflection on them, we presented the analyses to the entire administrative office, including representatives of various practitioner groups like municipalities or provinces.

After the presentation, reflections on our analyses proved helpful for theory testing and the creation of shared understanding. Being accompanied by our guide who had already shared an understanding with us (Coughlan and Coghlan, 2002) made it easier to organize support. A practitioner explained our findings framed in practitioner terms to other practitioners, thus effectively bridging the boundary between the social worlds of science and governance practice (Boezeman *et al.*, 2013; Clark *et al.*, 2011; Gieryn, 1995; Jasanoff, 1994). The reflection sessions are prime episodes of boundary work. In the same way, we reported back to the gatekeeper. Our analysis did not entail a comfortable message for the director of the administrative office, and therefore sharing knowledge was also about building trust and organizing support for acceptance within the social world of the administrative team. Again, our guide appeared crucial. During our final discussions with the gatekeeper, our guide explained the uncomfortable parts of the analyses in practitioner terms and, as our guide was an insider, this fitted well with the organizational understanding of the issue at stake. In addition, careful attention to the director's reflections on our analyses served to refine our analyses and created increased shared understanding.

Apart from powering our analyses into the administrative puzzle by giving presentations, reflecting, and creating a shared understanding, the analyses were disseminated throughout the organization to a large extent through the guide. Once the AR project finished, we as action researchers left the scene and lost our influence in the governance praxis. However, our guide continued to work in the administrative office and therefore could further disseminate our joint analyses. That the gatekeeper was responsible for much of the dissemination was also illustrated by a presentation that she gave at a climate adaptation conference a couple of months after the end of the AR project. She explicitly explained how the administrative office had learned from the AR project to give room to frame differences in the governance network to overcome apathy towards climate adaptation.

Analysis

This case illustrates how knowledge is seldom value-free in a constant praxis of frame interactions, as knowledge is constantly interpreted and employed in a process of sense making through framing. We have shown the way in which conducting effective AR has to deal with this complexity and how reflecting on the frames employed by the various practitioners created shared understanding of the governance landscape. In the case of the Dutch Delta Programme for Lake IJssel this understanding concerned the question of an initially productive process running out of steam. By reflecting with various stakeholders on the various frames employed, we finally created a shared understanding among stakeholders and DPIJ administrators on the differences in scales (local vs. national) and issues (technical vs. socio-economic) that had not been explicitly discussed before.

We were able to do so effectively only by constantly mapping the governance landscape and understanding the plurality of Wittgenstein's side-street views on a problem, which make a problem wicked. Instead of focusing on a linear main-street view (i.e. the official technical framing), careful mapping of the governance landscape not only led to better knowledge of the process, but also proved to be a strategic activity for organizing support by building a shared understanding of the analyses among the plurality of practitioners.

To map the governance landscape, legitimate access to the governance landscape was crucial and might have become a barrier if the network gatekeeper (Bache *et al.*, 1996) had not legitimized access. Building strategic relations with the gatekeeper of a governance network clearly enhances access to the actual governance praxis, and this helps to gain a more in-depth understanding of the puzzling and powering process in the network. Compared to, for example, document analysis or survey research, access to the actual governance praxis enabled the co-production of a shared understanding with practitioners.

In the wicked DPIJ context with various practitioners applying various frames to the climate adaptation issue, all having specific positions in the governance processes of puzzling and powering, finding one's way around as a newcomer could have become a second barrier. Partnering with a guide in the network proved extremely effective in this task. As we have shown, a guide should not only portray the official framing but also represent what Wittgenstein calls a certain *badness*. A 'bad' guide does not show the official framing, or the dominant view to a problem, but rather her or his natural habitat, with the strong advantage of knowing the various side-street views that constitute the authentic wickedness of a problem.

Accordingly, when the governance in wicked climate adaptation policy landscapes can be conceptualized as various side-street views or frames interacting as a process of puzzling and powering over climate adaptation, AR can be seen as a matter of 'powering to puzzle'. To introduce knowledge or reflect on governance praxis, AR is primarily a matter of organizing power to be able to take part in the collective puzzling. In addition, organizing power is needed to actually influence the puzzling process; this might be done by building strategic alliances and co-producing knowledge while participating in the governance praxis. Our guide was able to select important side-street views that we would otherwise have missed. In addition, our relation with the gatekeeper and our guide guaranteed access to these views and enabled us to coproduce knowledge with these views that could be disseminated in the network afterwards by both our guide and our gatekeeper.

The 'bad' guide facilitates the straddling of the demands from different social worlds (Boezeman *et al.*, 2013; Clark *et al.*, 2011; Gieryn, 1995; Jasanoff, 1994). Producing knowledge that impacts climate adaption governance does not appear to be a linear thing (Biesbroek *et al.*, 2010; Biesbroek *et al.*, 2013; Vink *et al.* 2013b). To understand the complexity of knowledge affecting governance, the wide variety of side-street views or frames in puzzling and powering over GACC should be taken into account. The way in which our guide made available the side-street views that were crucial to our AR project seems akin to the boundary work that Boezeman *et al.* (2013) describe as the success of the Dutch Delta Committee. As a boundary organization, the Delta Committee went beyond the classical definition of boundary work as demarcation work. Coordination work was important for its success, i.e. the way it positioned itself towards the various side-street views that it encountered in society and carefully collected. As the chairman of the committee phrased it: 'casting the net as widely as possible on the water', meaning that they collected a wide variety of side-street views on the issue of climate change. It facilitated a puzzling process in which the advisory report became attuned to the different issue framings on the one hand, while simultaneously negotiating the support of a network of powerful actors for whom the committee could credibly speak (Boezeman *et al.*, 2013).

Conclusion

We started this chapter on partnering with a guide in action research with the research questions: 1) how can partnering with a city guide as AR method help to gain a better understanding of the landscapes of actors, views, and positions that construct the wicked policy problem of adapting to climate change; 2) how can partnering with a city guide as AR method provide policy advice that fits the problem holders' authentic understandings of the problem and its surrounding governance landscape. Subsequently, we gave an account of how we participated in the governance network meetings with civil servants, decision makers, and stakeholders of the Lake IJssel Delta Programme after the programme director had asked us to advise on the process running out of steam. We described how partnering with a guide provided the possibility of in-depth frame reflections with the policy actors that would not have been possible in a less interactive research approach.

Accordingly, we discovered a difference in problem frames on the scale and nature of climate adaptation between public administrators organizing the network and the regional public and private actors participating in the network. By conducting frame reflections (Schön and Rein, 1994) through interviews and by actually taking part in the governance praxis through reflective meetings with a wide variety of practitioners, we were able to create not only an in-depth understanding of Wittgenstein's side-street views of the governance process, but also a shared understanding (Coughlan and Coghlan, 2002; Friedman and Rogers, 2009; Riordan, 1995) of the different scales and problem frames that had created the increased apathy among practitioners in the governance landscape. Partnering with a guide (Gasking and Jackson, 1967) who showed us the various side-street views and the network gatekeeper (Bache *et al.*, 1996) who provided access to the network, appeared crucial. Therefore we conclude that: (1) partnering with a guide can be an effective AR method to educe frame analysis of the various side-street views on climate adaptation and provide in-depth understandings of processes of puzzling and powering in climate adaptation governance; (2) liaising with a guide and the governance gatekeeper not only provided access to these various side-street views on governance, but also helped in the effective dissemination of the co-produced knowledge in the rest of the network; this can be seen as a special type of boundary work for effective and legitimate AR in climate adaptation governance.

Acknowledgements

We would like to thank the Dutch research programme Knowledge for Climate for providing the means to do this research. We would especially like to thank Ellen van Mulligen and Hetty Klavers of the Dutch Delta Programme for Lake IJssel for their dedicated, honest, and above all inspiring support as guide and gatekeeper.

Notes

1 Often these IJsselMeerdagen (IJsselMoredays/IJsselLakeDays; *more* and *lake* are the same word, *meer*, in Dutch) were attended by 80–150 participants, ranging from national civil servants and regional civil servants, to interest groups or occasionally a municipality council member. All actors were invited by the DPIJ administrative office
2 Decision-making conferences or Lake IJssel summits were attended by politically responsible decision makers such as county councils, majors, alderman, and water boards, often accompanied by their direct advisors and managers. Aside from one or two representatives of the private sector and societal groups, stakeholders and administrators were not invited. These conferences were attended by 40–120 decision makers and their direct advisors/managers.
3 Administrative (preparation) meetings were attended by regional public administrators, experts, and societal stakeholders and comprised 10–30 participants. All participants were invited by the administrative office.
4 The Hague is referring to the residence of the national government.

References

Bache, I. (2000) Government within governance: network steering in Yorkshire and the Humber, *Public Administration*, 78(3): 575–592.

Bache, I., S. George and R. Rhodes (1996) The European Union, cohesion policy, and subnational authorities in the United Kingdom, in L. Hooghe (ed.) *Cohesion policy and European integration*, 294–321, Oxford: Oxford University Press.

Barzilai Nahon, K. (2008) Toward a theory of network gatekeeping: a framework for exploring information control, *Journal of the American Society for Information Science and Technology*, 59(9): 1493–1512.

Benford, R.D. and D.A. Snow (2000) Framing processes and social movements: an overview and assessment, *Annual Review of Sociology*, 26(1): 611–639.

Biesbroek, G.R., J.E. Klostermann, C.J. Termeer and P. Kabat (2013) On the nature of barriers to climate change adaptation, *Regional Environmental Change*, 13(5): 1119–1129.

Biesbroek, G.R., R.J. Swart, T.R. Carter, C. Cowan, T. Henrichs, H. Mela, M.D. Morecroft and D. Rey (2010) Europe adapts to climate change: comparing national adaptation strategies, *Global Environmental Change*, 20(3): 440–450.

Boezeman, D., M. Vink and P. Leroy (2013) The Dutch Delta Committee as a boundary organisation, *Environmental Science & Policy*, 27: 162–171.

Buuren, A. van, M.J. Vink and J. Warner (in press) Constructing authoritative answers to a latent crisis? Strategies of puzzling, powering and framing in Dutch climate adaptation practices compared, *Journal of Comparative Policy Analysis*.

Checkland, P. and S. Holwell (1998) Action research: its nature and validity, *Systemic Practice and Action Research*, 11(1): 9–21.

Clark, W.C., T.P. Tomich, M. van Noordwijk, D. Guston, N.M. Dickson, D. Catacutan and E. McNie (2011) Boundary work for sustainable development: natural resource management at the Consultative Group on International Agricultural Research (CGIAR), *Proceedings of the National Academy of Science*, doi:10.1073/pnas.0900231108.

Coughlan, P. and D. Coghlan (2002) Action research for operations management, *International Journal of Operations & Production Management*, 11(1): 220–240.

Culpepper, P.D. (2002) Powering, puzzling, and 'pacting': the informational logic of negotiated reforms, *Journal of European Public Policy*, 9(5): 774–790.

Davoudi, S., J. Crawford and A. Mehmood (2009) *Planning for climate change: strategies for mitigation and adaptation for spatial planners*, London: Earthscan/James & James.

Delta Commissie (2008) *Working together with water: a living land builds for its future, findings of the Delta Commissie*, The Hague: Delta Commissie.

Dewulf, A. (2013) Contrasting frames in policy debates on climate change adaptation, *Wiley Interdisciplinary Reviews: Climate Change*, 4(4): 321–330.

Dewulf, A., B. Gray, L. Putnam, R. Lewicki, N. Aarts, R. Bouwen and C. van Woerkum (2009) Disentangling approaches to framing in conflict and negotiation research: a meta-paradigmatic perspective, *Human Relations*, 62(2): 155–193.

Entman, R.M. (1993) Framing: toward clarification of a fractured paradigm, *Journal of Communication*, 43(4): 51–58.

Ford, J.D. and L. Berrang-Ford (2011) *Climate change adaptation in developed nations: from theory to practice*, Dordrecht: Springer.

Ford, J.D., L. Berrang-Ford, A. Lesnikowski, M. Barrera and S.J. Heymann (2013) How to track climate change adaptation: a typology of approaches for national-level application, *Ecology and Society*, 18(3): 40.

Friedman, V.J. and T. Rogers (2009) There is nothing so theoretical as good action research, *Action Research*, 7(1): 31–47.

Gasking, D.A. and A.C. Jackson (1967) Wittgenstein as a teacher, in K.T. Fann *Ludwig Wittgenstein: the man and his philosophy*, 49–55, New York: Dell Puc. Co.

Gieryn, T.F. (1995) Boundaries of science, in S. Jasanoff, G. Markle, J. Petersen and T. Pinch (eds) *Handbook of science and technology studies*, 393–443, Thousand Oaks, CA: Sage.

Goodwin, M. and J. Grix (2011) Bringing structures back in: the 'governance narrative', the 'decentred approach' and 'asymmetrical network governance' in the education and sport policy communities, *Public Administration*, 89(2): 537–556.

Hall, P.A. (1993) Policy paradigms, social learning, and the state: the case of economic policymaking in Britain, *Comparative Politics*, 25(3): 275–296.

Heclo, H. (1974) *Modern social policy in Britain and Sweden: from relief to income maintenance*, New Haven, CT: Yale University Press.

Hisschemöller, M. and R. Hoppe (1995) Coping with intractable controversies: the case for problem structuring in policy design and analysis, *Knowledge and Policy*, 8(4): 40–60.

Hovi, J., D.F. Sprinz and A. Underdal (2009) Implementing long-term climate policy: time inconsistency, domestic politics, international anarchy, *Global Environmental Politics*, 9(3): 20–39.

Hulme, M. (2009) *Why we disagree about climate change*, Cambridge: Cambridge University Press.

Jasanoff, S. (1994) *The fifth branch: science advisers as policymakers*, Cambridge, MA: Harvard University Press.

Jordan, A., D. Huitema, H. van Asselt, T. Rayner and F. Berkhout (2010) *Climate change policy in the European Union: confronting the dilemmas of mitigation and adaptation*, Cambridge: Cambridge University Press.

Keskitalo, E.C.H. (2010) *Developing adaptation policy and practice in Europe: multi-level governance of climate change*, Dordrecht: Springer.

Lazarus, R.J. (2008) Super wicked problems and climate change: restraining the present to liberate the future, *Cornell Law Review*, 94: 1153–1233.

Lempert, R., J. Scheffran and D.F. Sprinz (2009) Methods for long-term environmental policy challenges, *Global Environmental Politics*, 9(3): 106–133.

Magnani, L. (2001) *Abduction, reason, and science: processes of discovery and explanation*, New York: Kluwer Academic/Plenum Publishers.

Phillips, N. and C. Hardy (2002) *Discourse analysis: investigating processes of social construction*, Thousand Oaks, CA: Sage.

Repetto, R.C. (2008) *The climate crisis and the adaptation myth*, New Haven, CT: Yale School of Forestry & Environmental Studies.

Rhodes, R.A.W. (1996) The new governance: governing without government, *Political Studies*, 44(4): 652–667.

Riordan, P. (1995) The philosophy of action science, *Journal of Managerial Psychology*, 10(6): 6–13.

Rittel, H.W.J. and M.M. Webber (1972) *Dilemmas in a general theory of planning*, Berkeley, CA: University of California Press.

Schön, D.A. and M. Rein (1994) *Frame reflection: toward the resolution of intractable policy controversies*, New York: Basic Books.

Stoker, G. (1998) Governance as theory: five propositions, *International Social Science Journal*, 50(155): 17–28.

Stone, D.A. (1989) Causal stories and the formation of policy agendas, *Political Science Quarterly*, 104: 281–300.

Termeer, C., A. Dewulf and G. Breeman (2013) Governance of wicked climate adaptation problems, in J. Knieling and W.L. Filho (eds) *Climate change governance*, 27–39, Berlin: Springer.

Vink, M.J. and E. Mulligen (2013) *Evaluatie lerend proces Deltaprogramma IJsselmeergebied*, Lelystad: Deltaprogramma IJsselmeergebied.

Vink, M.J., D. Boezeman, A. Dewulf and C.J.A.M. Termeer (2013a) Changing climate, changing frames: Dutch water policy frame developments in the context of a rise and fall of attention to climate change, *Environmental Science and Policy*, 30: 90–101.

Vink, M.J., A. Dewulf and C. Termeer (2013b) The role of knowledge and power in climate change adaptation governance: a systematic literature review, *Ecology and Society*, 18(4): 46.

Wagenaar, H. (2011) *Meaning in action*, Armonk, NY: ME Sharpe.

Warner, J., P. Wester, M.J. Vink and A. Dewulf (in press) The politics of framing scales, ambiguity and uncertainty: flood interventions in the Netherlands, in A. Cohen and E. Watson (eds) *Negotiating water governance*, London: Ashgate.

Weick, K. E., K. M. Sutcliffe and D. Obstfeld (2005) Organizing and the process of sensemaking, *Organization Science*, 16(4): 409–421.

Wood, L.A. and R.O. Kroger (2000) *Doing discourse analysis: methods for studying action in talk and text*, Thousand Oaks, CA: Sage.

Yanow, D. (1996) *How does a policy mean? Interpreting policy and organizational actions*, Washington, DC: Georgetown University Press.

Yanow, D. and P. Schwartz-Shea (2006) *Interpretation and method: empirical research methods and the interpretive turn*, Armonk, NY: ME Sharpe.

4 Participation in climate adaptation in the Lower Vam Co River Basin in Vietnam[1]

Patrick Huntjens, Bouke Ottow and Ralph Lasage

Introduction: a climate change adaptation strategy in Long An province, Vietnam

This chapter describes a participatory planning process to develop a climate change adaptation strategy for the Lower Vam Co River Basin in Long An province in Vietnam under an 18-month pilot project (September 2011 to April 2013). Given the Vietnamese culture and context, this could not be done in the same way as in Western cultures. This chapter addresses the extent to which action research methods can be applied in a non-Western culture such as Vietnam, taking into account cultural differences and possible ways to bridge these. We start with a problem statement, followed by a description of the kind of action research applied and the empirical study conducted.

People living in the villages and cities of Long An, as well as socio-economic development and ecosystems in this region, are under serious threat from the impacts of climate change. The pilot area covers four southern coastal districts where floods from upstream combined with high tides from East Sea (via the Soai Rap estuary) may lead to severe inundation. In addition, drought and salinity intrusion, especially in the dry season, are already causing serious problems for freshwater supplies for domestic, agricultural, and industrial uses in the area. This problem is further aggravated by dwindling groundwater resources due to over-exploitation. Under present conditions, climate variability is already influencing successful water management. The impacts of climate change increase the existing complexities of achieving just socio-economic and sustainable development involving multiple uses of water among growing numbers of users. In summary, climate change has severe negative consequences for water resources, human health and safety, food production, industry, and navigation in the pilot area.

Group model building in participatory planning

The participative planning process in this pilot project involved local actors at different governance levels (i.e. provincial, district, and commune level) to ensure inclusion of their priorities and the challenges they face in reducing climate risks. The project design was built on the assumption that all stakeholders have relevant experience, knowledge, and information that ultimately will inform and improve the quality of the planning process as well as any actions that (may) result. Participation processes and local knowledge play a crucial role in the implementation of adaptive management (AM) in river basins (Alkan *et al.*, 2006; Reed, 2008). According to Reed (2008), participation should be underpinned by equity, trust, and learning to be effective and have an impact. We consider these elements important for stakeholder processes in water management, which in general aim at creating ownership and awareness of the different views and perceptions that exist on a problem in a basin (Pahl-Wostl, 2007). Accordingly, they also aim at initiating social learning while engaging local and regional stakeholders to bridge the science–policy gap and to improve the practical relevance of research (cf. experiences of Alkan *et al.*, 2006; learning alliances as reported in van Buuren and Edelenbos, 2004; Vries, 2006; review by Reed, 2008). Via beneficial collaboration, both scientists and consultants (hereafter called action researchers) – in our case ecologists, hydrologists, sociologists, governance specialists, and planners from the pilot project 'Participation in Climate Change Adaptation' – and relevant stakeholders – in our case farmers, fishermen, professionals in water management, representatives of the administration, academics, private-sector and NGO representatives – come closer to the nature of the problems in a basin (Hart, 1986; Hodgson, 1992; von Korff, 2005; Pahl-Wostl, 2007). We see early and broad stakeholder participation as a prerequisite for putting AM into operation.

A wide range of very different methodologies exist for initiating, enhancing, facilitating, and supporting participation and stakeholder processes. Within the setting of this pilot project in Vietnam, we concentrate on a specific method of action research: group model building (GMB). GMB is increasingly attracting attention in the field of complex decision making, public policy making, and implementation (Cockerill *et al.*, 2006, 2007; Rouwette *et al.*, 2000; Vennix, 1999; Zagonel and Rohrbaugh, 2007) and in particular as a useful method for stakeholder participation in water management to develop and improve decision support models as well as integrate existing information with local/stakeholder knowledge (den Exter, 2004; Hare *et al.*, 2006; Stave, 2002).

GMB is a method for facilitating the deep involvement of a group of individuals in the building of a model of a particular management system. The objective is to improve group understanding about that system, its problems, and possible solutions, in the expectation that this may directly or indirectly lead to better management decisions (Hare, 2003; Vennix, 1999). During the pilot project, GMB helped researchers and stakeholders to look systematically at the integration of different knowledge frames, conflicting attitudes, and ideas of what is wanted and needed.

GMB (also called cooperative modelling, collaborative modelling, participatory modelling, mediated modelling) can be described as a collection of pieces of a facilitated group exercise and of techniques used to construct joint kind-of-model representations of the system that move a group forward in a systems thinking intervention (according to Andersen and Richardson, 1997). During

(a)

(b)

Figure 4.1 Group model building in break-out groups during a district meeting

GMB, the initiator (or facilitator) of the participative process works directly with stakeholder groups on key problems/(shared) visions (Palmer, 1999) or decisions. The method assumes that the knowledge available among the participants can help jointly work out key elements and relationships of the problem at stake more effectively. GMB exercises make the different mental models of the participants explicit and may confront all participants with them. Although the ideal result of a GMB process is to bring these different mental models up to a synthesis, this happens only rarely. Many authors (e.g. den Exter, 2004; Hare *et al.*, 2006; Stave, 2002) show that system dynamics GMB, in particular, has the potential to facilitate stakeholder learning and assist stakeholders to think holistically about the complex systems they are trying to manage.

Public participation in the Vietnamese culture

Various studies and practical experiences in the Netherlands and elsewhere show the importance of active stakeholder involvement in water management (Huntjens *et al.*, 2011, 2012; von Korff, 2005; Pahl-Wostl, 2007). From the literature it is also known that the direct use or application of Western concepts, methods, and approaches in other cultures is not always successful, e.g. Arnstein's acclaimed participation ladder has resulted in misleading results in a development context (Hostovsky and Maclaren, 2005). Regarding participation in environmental impact assessments (EIAs), Hostovsky and Maclaren argue for more culture- and context-sensitive EIAs in Asia, rather than a slavish imitation of Western EIA paradigms. The role of NGOs in Vietnam, for example, is not comparable to their role in the Netherlands: Kerkvliet (2002, quoted in Hostovsky and Maclaren, 2005) finds that it is possible for Vietnamese citizens to express their concerns, but only through organizations that are authorized and financed by the state. In this way, the Communist Party mobilizes public support, whilst maintaining control.

In order to develop a culture-sensitive EIA or other participatory process, as advocated by Hostovsky and Mclaren (2005), it is important to know the local culture. According to Duong (2004), Confucian traditions largely determine the behaviour and relationships between Vietnamese people. The interests of society stand above those of the family and individuals, although the rights of individuals are respected, provided that they are not opposed to those of the family, the village, and the country (Duong, 2004).

Duong's (2004) observations are in line with Hofstede *et al.*'s (2010) culture classification. Based on studies in 76 countries, their classification defines four culture descriptive indices: power distance, individualism, masculinity, and uncertainty avoidance (Figures 4.2 and 4.3).

In Vietnam, power distance is high (22nd out of 76 countries), whereas it is low in the Anglo and northwest European countries (around place 60). These countries score very high on individualism (1 to 19), whereas this score is very low for Vietnam (58th). The differences in masculinity and uncertainty avoidance are (much) smaller, and in these indices the differences between the Anglo and

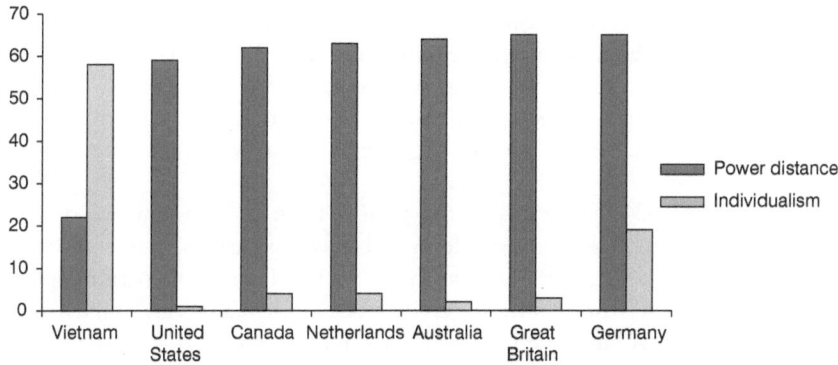

Figure 4.2 The relative scores of Vietnam and northwest European and Anglo countries on Hofstede's cultural indices 'power distance' and 'individualism'

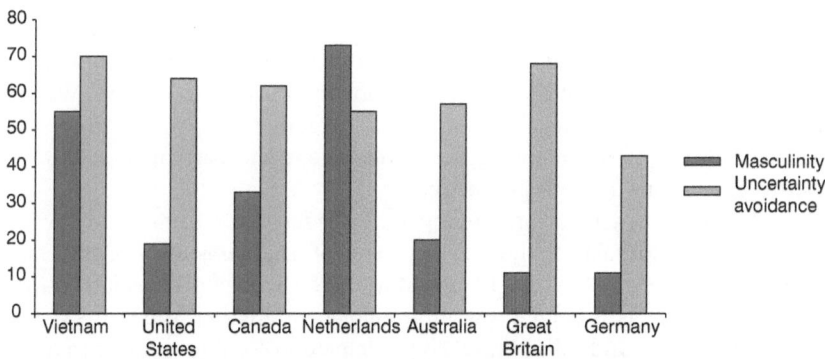

Figure 4.3 The relative scores of Vietnam and northwest European and Anglo countries on Hofstede's cultural indices 'masculinity' and 'uncertainty avoidance'

northwest European countries is also significant, for instance on masculinity between the Netherlands and the other countries and on uncertainty avoidance between Germany and the other countries.

The cultural indices entail some important cultural characteristics that are relevant to public participation. For example, a high power distance (as in Vietnam) is characterized by hierarchy in organizations, centralization and formal rules, confidence by subordinates in their managers, and the expectation that they will be told what to do. Important features associated with a low individualism (as in Vietnam) include: different standards for in-groups and out-groups, exclusionism, striving for harmony, and avoidance of direct confrontations. These aspects of Vietnamese society may hinder participation in decision making by individual citizens, as we advocate in the Netherlands and the West generally.

It is important to note the low scores for Vietnam on masculinity and uncertainty avoidance. One of the features of a society with a low score for masculinity is 'resolving conflicts through compromise and negotiation'. For uncertainty avoidance, it is important to mention that, according to Hofstede *et al.* (2010), people consider uncertainty and unknown risks a normal aspect of life; in addition to being comfortable with ambiguous situations and unknown risks, they quickly accept new things such as mobile phones, email, and the Internet, citizens are competent in relation to authorities, civil protest is acceptable, and there is a high degree of participation in voluntary organizations and movements.

From the shared experiences of the project team in Vietnam, we do not recognize aspects such as 'comfortable in ambiguous situations' and 'competent citizens in relation to authorities'. On the other hand, we do recognize aspects such as 'quickly accept new business', 'civil protest is acceptable', and 'high degree of participation in voluntary organizations and movements'. These latter aspects can be helpful for a participatory process.

Doi Moi and Grassroots Democracy in Vietnam

In Vietnam, water-user and/or farmer associations are becoming increasingly influential, where the emergence of community-based organizations is linked to a set of policy reforms (Doi Moi, literally 'renovation') starting in 1986. Within this context of Doi Moi, a legal framework for integrated water management was established in the 1990s, allowing the development and participation of civil society organizations in the water sector.

In 1997, various abuses surrounding local leaders led to unrest in several provinces. This stimulated the development of the Grassroots Democracy Decree (Decree 29/1998/ND-CP) in May 1998, with further elaborations in 2003 and 2007. This decree was an attempt to strengthen the rights of the people at commune and village level to participate in local government affairs. An important aspect remains the socialist ideology of collective supremacy in a centralized democracy, namely, the views of people at local level do not directly reach the central government, but are summarized and reported up through the chain of command (Duong, 2004).

People take part in democracy by voting for members in the local people's council and the National Assembly, by participating in local government affairs as laid down in the Grassroots Democracy legislation, and by directly sending comments to members of the National Assembly. Another way is through participation in social organizations affiliated to the Communist Party, such as the women's and youth organizations (Duong, 2004).

The Grassroots Democracy Decree distinguishes four categories of cases in which people have different rights, based on an old slogan from the beginning of the struggle for independence:

- *dân biết*: people know (i.e. work/activities to be informed about);
- *dân bàn*: people discuss (i.e. work/activities to be directly discussed);
- *dân làm* people do (i.e. work/activities to be consulted about); and
- *dân kiểm tra*: people monitor (i.e. work/activities to monitor or inspect).

The Grassroots Democracy legislation indicates, by category, examples and methods that can be used (Duong, 2004). Examples of matters about which the public must be consulted are land-use planning and the development of strategies to achieve national targets, such as a climate change adaptation strategy at provincial level. Methods to consult the public on these matters, in addition to meetings, include surveys to collect opinions.

In the years following the introduction of the Grassroots Democracy Decree, researchers have found positive effects, but the effects vary in the different areas studied (Duc and Minh, 2008; Duong, 2004; Mekong Economics, 2006). A recent analysis of water governance in Vietnam (Huntjens and Bossert, 2010) still observes a strong hierarchy in water, limited social integration, and very limited access to information for stakeholders.

Project methodology

Development process

The methodology for this pilot project was initially developed by the action researchers. After approval and 80 per cent co-funding by the Dutch Government (in September 2011), the action researchers started with the initiation phase. This phase included a consultation process with key stakeholders[2] in the pilot area, i.e. provincial government departments[3] and two research institutes from Ho Chi Minh City, to develop a methodology tailored to their needs and their cultural setting. An important outcome of the initiation phase was a concept methodology, which was discussed with a larger group of stakeholders in a consultation meeting in Long An Province (in January 2012). This concept methodology was included in a proposal to be submitted for approval by the Long An Provincial People's Committee (PPC Long An). In March 2012, PPC Long An approved the project, including a commitment of 20 per cent co-funding, after which the first provincial multi-stakeholder dialogues were organized in March 2012.

During implementation of the project, we were confronted with a number of challenges:

- Whereas the project design assumed that the necessary information from other government organizations through the counterpart (government) organization would become freely available, this information was not easy to get and had to be paid for separately by the project.
- It was not possible to sit around the table with our counterparts to make a stakeholder analysis. First, it was not well understood why such an analysis was needed. Second, our Vietnamese counterparts insisted that they could not say anything about the position and interests of other agencies involved with water and climate in the Lower Vam Co Basin. We consider this, as also the first point, an expression of a hierarchical culture, in which communication is primarily vertical.
- Participants in the training sessions prior to the planned stakeholder consultation meetings were not a selection of people who would play a role

in the organization and facilitation of this, but 'just' all attendees for the consultation meetings. This reduced the effectiveness of the training focused on facilitation, participation, and decision support methods. This might very well be an aspect of the low-individualistic/high collectivistic culture.

Because some methods, in particular GMB, had never before been used in the project area, the action researchers decided that the planning process should entail a certain degree of methodological flexibility, by offering an iterative step-by-step approach for strategy development. This iterative process was made possible by providing and collecting evaluation sheets from participants, as well as evaluation by the project team, after finalizing each step of the process. Thus, lessons learnt from previous steps were taken into account in designing the next steps, both in terms of process management and methods used. Additionally, an external advisory committee[4] was established for the purpose of monitoring and evaluation. This committee met three times in total after key milestones (i.e. multi-stakeholder dialogues) of the planning process. The GMB workshops at provincial and district level were part of the dialogue track.

Group model building process

It is important to consider the purpose of involving participants from science and practice (as explained above) in the GMB process. The variety of techniques used for GMB relates to its many purposes and levels of participation (den Exter, 2004). Goals may include encouraging team (social) learning and communication, and improving the chances of the goals of a specific (policy or management) process being achieved by (1) improving inputs from expert knowledge, (2) improving the uptake and commitment of the participants to the results of the process, and (3) improving the democratic accountability of the management process. From a scientific point of view, GMB has been used to involve multiple experts to increase the validity and robustness of an existing conceptual, or already coded computational, model on ecosystem or water management functioning (such as presented by Costanza *et al.*, 1990, and Costanza and Ruth, 1998).

Stakeholders may participate in GMB at different stages. Stave (2002) identifies a spectrum of participation when stakeholders are being involved in model building. At the least participatory level, water managers can use a completed model to demonstrate the effects of alternative policies to other stakeholders. Depending on the way the discussion is facilitated, this approach can greatly enhance participants' understanding of the resource system and the effects of alternative management decisions. A more participatory scenario would allow stakeholders from different fields to suggest their own strategies to be tested (by a model; cf. Costanza and Ruth, 1998). At the most participatory level, stakeholders develop their own – preferably shared – conceptual model that represents their view of the system structure, as reported by Cockerill *et al.*, (2006). The latter level was targeted in the GMB process described in this chapter.

Response of the pilot project to the cultural setting

In response to above observations, and the challenges identified in the second section, the project team came to the following conclusions during the initiation phase of the project (September 2011–March 2012):

- participation does have a base in the culture of Vietnam and fits into recent developments (i.e. Doi Moi policy), but in a different way than we are used to in the Netherlands, e.g. in Vietnam representation takes place via mass organizations;
- developing a climate adaptation plan fits into the Grassroots Democracy category of cases in relation to which citizens should be consulted;
- we must take into account the existing hierarchy;
- the stakeholder analysis, which initially we could not do well, should be gradually developed by means of questionnaire surveys (an accepted method) during workshops at provincial and district level, in which respondents only say something about themselves and their organizations;
- it is important to involve the levels below the province in the process: the districts and communes, and if possible also the villages – this would require district meetings to be organized and surveys to take place at village level;
- the mass organizations (e.g. farmers, women, youth) should certainly be involved in the process;
- existing NGOs should be fully involved in the process; and
- interactive forms are perhaps not common, but can be used if properly prepared, initiated, and facilitated.

The proposed multi-stakeholder process was shaped by a number of multi-stakeholder meetings at provincial and district level, and surveys at commune/village level. The dialogues were facilitated using advanced methods of participation, such as cognitive mapping and group model building, accompanied by adequate information at the right time during the planning process. The latter was achieved by providing the outputs of risk assessment and hydrological modelling when necessary. We give a brief description below of the process and the methods used.

Multi-stakeholder dialogues: provincial level

Multi-stakeholder dialogues were the backbone of the participation process in the project. At two large provincial meetings in March and May 2012, we established a joint problem definition with the stakeholders, and identified potential solutions and strategy components, as well as criteria for the assessment of these strategy components, parallel and interacting with the hydrological modelling (2D SOBEK). These meetings were attended by 32 representatives from 18 provincial organizations and nine representatives from three different district-level organizations. These were mainly ministries and people's committees, a number of mass organizations, but unfortunately no NGOs.

District meetings

Between the first and second provincial meeting, multi-stakeholder meetings were held in each of the four districts. During these district meetings, problems and possible solutions were explored. In total, there were about 100 representatives from 32 different district-level organizations and also about 100 participants who represented the commune level, including two NGOs. During the district meetings, participants also filled out a form about the position, interests, role, and goals of their organization.

Training

Prior to the first provincial stakeholder meeting and prior to the district meetings, we had two days of training in which participation and decision support principles and surveying principles were discussed. During the first training session, we asked and facilitated the participants to conduct a stakeholder analysis for the project. This analysis was used to determine the persons to be surveyed (see next section). Additionally, the session prepared Vietnamese participants to support the multi-stakeholder dialogues, e.g. by facilitating break-out groups themselves.

Surveys at village level

On the training day prior to the district meetings, we asked participants to determine which persons should be involved in the survey on issues surrounding water management and climate change and possible solutions. Eventually, and based on the above preparatory steps, 120 people were surveyed at commune level.

Group model building

At the first provincial multi-stakeholder dialogue, as well as in the four district meetings, we used GMB in small groups, i.e. four group models in each district, thus 16 in total at district level. The purpose was to come to a joint problem definition, to identify possible solutions, and to identify priorities. Additionally, participants were asked to position the prioritized measures on a land use map of their own district (see Figure 4.4).

The participants started hesitantly at the beginning of the workshops, but their hesitation disappeared soon after some persons took the initiative. In the GMB sessions, a 'round robin'-type method helped people to overcome the barrier to saying something in the group. Feedback evaluation forms indicated that the participants valued the integrative aspect of the method and the possibility of integrating their own ideas in a structured way.

In order to develop targeted policy recommendations, the solutions were assessed in direct connection to the joint problem definition by means of group

Figure 4.4 During the district meetings participants were asked to position the prioritized measures on a land use map of their own district

model building. This means that specific solutions were linked to specific problems. They used a synthesized group model of problems and solutions relating to sustainable and climate-resilient water management in the Lower Vam Co Basin in Long An Province. The synthesis is based on the results of 18 group models developed in the provincial and district meetings in March and April 2012. The group model uses a holistic and summarized overview of the perspectives of all stakeholders in the pilot area.

Building strategy components

From the outcome of the models and the stakeholder meetings, the project team prepared five optional strategy components (see Table 4.1) to sketch the playing field for strategy development. These strategy components were discussed during the multi-stakeholder dialogue on 23–24 May 2012 in Tan An, Long An Province. The details and feasibility of the five strategy components were evaluated by the participants during the design sessions, supported by experts. During the design sessions, the participants were divided in small groups in which they evaluated each strategy component on a given set of criteria.

Table 4.1 Overview of strategy components, interventions sheets, and type of intervention for different strategies

Strategy	Strategy component	Intervention	Type of intervention
1. No regret	Upgrading existing dyke system	1a. Upgrading existing dyke system	Physical intervention
		1b. Construction of sluices	Physical intervention
	Optimizing water supply and water demand measures	1c. Rainwater harvesting	Physical intervention
		1d. Surface water extraction	Feasibility study
		1e. Development of sustainable groundwater management policy	Governance intervention
		1f. Water transfers within pilot area	Feasibility study
		1g. Desalination	Feasibility study
		1h. Point-of-use conservation	Physical intervention
		1i. Water-saving technologies in irrigation	Physical intervention
		1j. Land use change from agriculture to aquaculture	Governance intervention
		1k. Water recycling and re-use	Physical intervention
	Upgrading natural water capacity	1l. Dyke replacements	Feasibility study
		1m. Increasing retention capacity	Feasibility study
		1n. Wetlands development and/or restoration	Governance intervention
	Improving governance tools	1o. Economic and financial measures	Governance intervention
		1p. Communication and education measures	Governance intervention
		1q. Regulatory measures	Governance intervention
2. Tidal barrier at Vam Co mouth	Construction of infrastructure	2a. Construction of tidal barrier	Feasibility study

Strategy	Strategy component	Intervention	Type of intervention
3. Import of freshwater	Import of freshwater from own river basin	3a. Import of freshwater from upstream groundwater	Feasibility study
		3b. Import of fresh water from upstream surface water	Feasibility study
	Import of fresh water from other river basins	3c. Import of fresh water from adjacent river basins	Feasibility study

Multi-criteria analysis

To be able to assess the effects of the different strategy components under different climate change scenarios, multiple criteria were used. These criteria enabled us to get insight into the effects of the strategy components on the livelihoods of the people in Long An province. They linked the physical characteristics of the system and the different land uses and sectors in the province.

In this analysis, we included three strategy components focused on flood safety and water availability, and two strategy components aimed solely at water availability and water demand. In the latter, we had to exclude investment costs, as these were not available. The information that was available was included in the model and analysis, together with expert judgment. We weighted the criteria on the basis of the involved stakeholders' perceptions.

Preferred strategy

The evaluation of these five strategy components resulted in the project team developing three integrated strategies focusing on continuing current land use and economic activities:

1 no-regret strategy
2 tidal barrier and sluices
3 import freshwater

Table 4.2 gives an overview of the integrated strategies (including related strategy components, intervention sheets, and type of intervention for different strategies), in which strategies 2 and 3 are additions to the no-regret strategy (strategy 1). Each strategy includes a package of measures for each strategy component and intervention sheets for each measure (name, location, objective, responsibility, impact, and so forth). Overall, 21 detailed intervention sheets were developed.

Table 4.2 Overview of strategy components

Strategy component 1 *Completing dyke system and upgrading dykes* Upgrading existing dyke system • Repair weak spots • Increase dyke level to +2m mean sea level Construction of dyke rings • Dyke rings comprise sluice gates (12 in total) • Gates can be closed in case of flood – closed ring • Safety level depends on maintenance, type of construction, interface between gates and dyke itself Relocate dykes in upstream part of pilot area	
Strategy component 2 *Semi-permeable barrier at river mouth + sluices to stop saltwater intrusion at high tide* All measures of strategy component 1 Additional is a tidal barrier in Vam Co river • Shipping problems may arise • In dry season already minimized depth • Better protection against high water levels of the sea	◆ Existing sluice ◆ Planned sluice (approved ot not yet apporved) Tidal barrier
Strategy component 3 *More room for rivers and nature – to increase resilience and water discharge capacity of river system* Retention areas for mitigating peak discharges in wet season, and storage of water for freshwater supply in dry season New areas outside dykes (created by dyke replacements) might be used as wetland areas, mangrove reforestation and/ or exploitation (in particular downstream areas), or cajaput forests (in particular upstream areas) Extension of existing wetlands	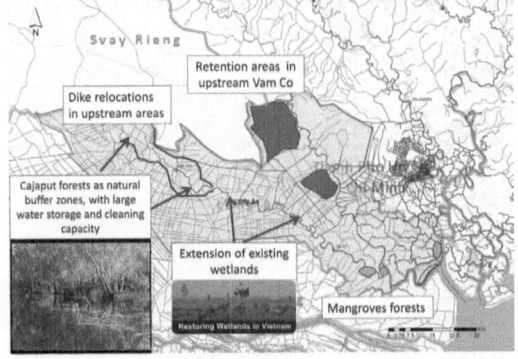

Strategy component 4
Import freshwater
Transfer of additional freshwater or groundwater from neighboring districts of provinces – this requires exchange arrangements
Possible sources
- Mekong river
- Saigon river, via Rach Tra diversion (Dau Tieng Reservoir)

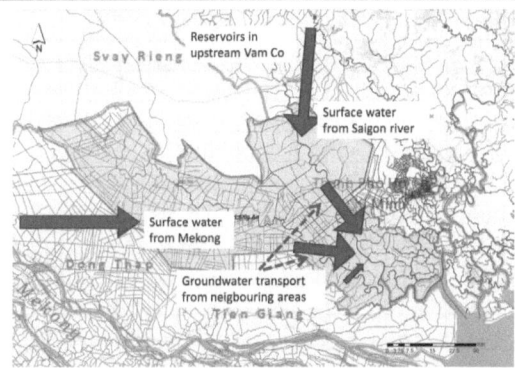

Strategy component 5
Self-sufficiency in fresh water supply
Optimize supply-side options in pilot area
- Rainwater
- Surface water
- Ground water
Optimize demand-side options in pilot area
- Point-of-use conservation
- Efficient irrigation systems
- Land use changes – from irrigation to aquaculture
- Water reuse and recycling

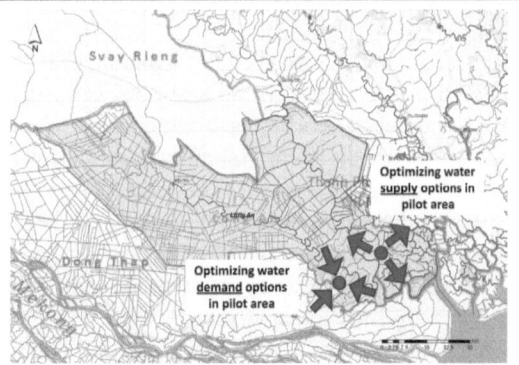

Discussion and conclusion

As evaluated by the External Advisory Committee (EAC) and the panel discussions during the mid-term and final conference, the existing Vietnamese demand for adaptive and participatory approaches to water management and water governance has been served sufficiently well by the project, and provides a good basis and reference for further dissemination and upscaling. Additionally, using the evaluation forms distributed and collected during the multi-stakeholder dialogues at provincial and district level, the project team was able to monitor the project, resulting in an evaluation of the participative planning process and action research methods being used.

Judging performance, or the effectiveness of multi-stakeholder dialogues and participative planning more in general, is challenging. Identification and attribution of specific outcomes is often confounded by other social and political processes that surround dialogue interactions. Nevertheless, some of the methods for assessing public participation methods more broadly are likely to be useful. One approach is to consider acceptance and process criteria (Rowe and Frewer, 2000). Acceptance criteria that have been suggested include representativeness, independence, early involvement, influence on policy, and transparency. Process criteria include access to resources like information, clear task definition, structured decision making, and cost effectiveness. For dialogues where the objective is to elicit consideration of alternatives, other criteria relating to being visionary and deliberative should also be considered (Dore, 2007).

In general, with a combination of action research methods (i.e. GMB, cognitive mapping, and multi-stakeholder dialogues) and advanced decision support tools, the pilot project demonstrated the added value of applying a participative planning process for the integration of different interests and spatial challenges into one single strategy. The pilot project shows that participatory planning methods, such as group model building and cognitive mapping, can be applied in a fruitful way, if properly embedded, initiated, and facilitated. The following paragraphs focus on content and format in order to reflect upon what is needed to achieve this.

First, the knowledge content of a multi-stakeholder dialogue is critical – what topics and issues are covered and how well informed the debate about them is. This often depends on access to scientific and experienced-based knowledge. At the same time, deliberative opportunities – time to question, seek clarification, discuss assumptions, and examine arguments – are thought to be critical for dealing with contested knowledge claims and exploring alternatives and poorly known risks and interests.

In this pilot project, an effective planning process was achieved by providing adequate information and inviting key stakeholders to participate at the right stage of the strategy development process. The latter was achieved by providing the outputs of risk assessment and hydrological modelling whenever necessary, e.g. before evaluation of the strategy components during the second multi-stakeholder dialogue. Adequate information supply was made possible thanks to a variety of decision support tools, such as the SOBEK modelling, a GIS-based damage module for the pilot area, and DEFINITE for the multi-criteria analysis.

The quality of participation is a function of many factors, including venues, session formats, how agendas are set, time and quality of briefing materials, and facilitation. It is still common practice for governments to sell pre-defined plans and call it participation or consultation (sometimes called a 'decide and defend' approach. Facilitators and conveners have crucial roles in determining the meaningfulness of participation and depth of deliberation (Huntjens, 2011; Lebel *et al.*, 2011).

Second, an important format issue is how facilitation, meeting structure, and venues influence the openness and multi-directionality of conversations that can take place. In practice, there are many different ways in which interaction between stakeholders or between public and government can be managed (Huitema *et al.*, 2009; Rowe and Frewer, 2000). In this pilot project, the group model building improved group understanding about the water management system, its problems, and possible solutions; this will directly or indirectly lead to better management decisions. Not only is the model itself a product of the process, the generation of common understanding among the stakeholders during the process is another important product of the process. Overall, stakeholders indicated that they appreciated the integrated and holistic approach and the way their own ideas were incorporated in a structured way.

For implementing a participative planning process, including action research methods, in Vietnam, the following *important lessons* are drawn:

- Characteristics of Vietnamese society, such as conflict resolution by negotiation, an open attitude to new technologies and innovations, and the high level of participation in voluntary mass organizations, constitute a good basis for participation in decision making and planning.
- The active involvement of mass organizations and NGOs through which citizens' interests are represented seems to be an effective approach for participative planning in Vietnam. Mass organizations play an important role in organizing people at local levels (district, commune) and at grass-roots level, amongst others, because the number of people belonging to organizations is very high in Vietnam compared with countries like Singapore and China, which have comparable types of governance. The five mass organizations (or socio-political organizations) in Vietnam include the Women's Union, the trade unions, the Youth Union, the Farmers' Association, and the Association of Veterans. The Fatherland Front, which is the umbrella organization of the mass organizations and other organizations, is also counted as a mass organization.
- From our observations during the district meetings and from interviews with village heads in the project area, on the lowest government level (grass-roots level), the village head seems to play an important role. As highest in rank on behalf of the local government, and given the existing hierarchy, the village head enjoys wide respect among (at least a large part of) the local community. At the same, the village head is expected to have a good overview of local interests and to represent these interests at the next governmental levels (i.e. district and province). Thus, the village head is considered a crucial link between local affairs and higher levels of government.
- The reform policy (Doi Moi) since 1986 and the recent development in Grassroots Democracy policy have laid a good foundation for a participatory planning approach in Vietnam. The Grassroots Democracy legislation indicates, by category, examples and methods that can be used for participation, e.g. via mass organizations (Duong, 2004).
- Political buy-in and ownership by the provincial government (e.g. via co-funding of the project) was considered an important success factor for the institutional embedding of the project's process and outcomes. The co-funding by the PPC required a justification of public resources well spent. Hence, the involved government departments (in particular the Department of Science and Technology, Department of Agriculture and Rural Development, and the Department of Natural Resources and Environment) were all held responsible by the PPC for successful completion of the project.

The project followed a *multi-level governance approach* to support integration between different levels and to establish liaisons and opportunities for upscaling the project results. The pilot project organized multi-stakeholder dialogues at three different levels (province, district, and commune), and the EAC additionally supported interaction and collaboration with higher levels, including regional, national, and international. The EAC members were specifically selected to allow for upscaling and streamlining of the project results. For example, the no-regret strategy consists of a consensus-based mix of measures that can be implemented safely without compromising the development of wider water management plans, such as the Mekong Delta Plan or the Ho Chi Minh City Flood Management Programme. The no-regret strategy aims at increasing dyke height, creating more room for the river, protection of nature areas (mangrove swamps, and so forth), and optimizing local supply and demand management.

In this pilot project, the action researchers brought relevant stakeholders into contact with one another. As a direct result, they managed to enhance levels of trust between the different actors, to share information and knowledge, and to generate solutions and relevant good practices. The resulting climate change adaptation preferred by the stakeholders is based on problems and solutions identified by an extensive group of stakeholders within the pilot area and has been developed on the basis of expert judgement, action research methods, and advanced decision support modelling tools.

The complexity and sensitivity of water and climate issues demand well-designed strategy components and related interventions that can operate successfully within relevant cultural, political, and economic settings. Accordingly, the processes relevant to designing and implementing these interventions were considered as important as the outcomes (e.g. the 'how' can matter more than the 'what'). In this respect, the political buy-in and ownership by the provincial government was crucial. Moreover, the people or organization(s) chosen to lead and participate in an intervention are deemed critical to gaining or losing political buy-in and sustainability.

This is one of the first projects in Vietnam to introduce a comprehensive approach for full-scale and meaningful participation of relevant stakeholders at different levels (province, district, commune) to develop a climate change adaptation strategy. In Vietnam, this is unique because the participative planning process covers a complex array of problems (i.e. floods, salt water intrusion, water scarcity, droughts, and water pollution) and multiple spatial challenges in one area.

The project conducted an extensive monitoring and evaluation procedure on the participative planning process, based amongst other things on the evaluation forms during the multi-stakeholder dialogues at provincial and district level, feedback from participants during the midterm and final conference, and feedback from the EAC. Overall, stakeholders indicated that they appreciated the integrated approach and the way their own ideas were incorporated in a structured way. During the final conference, we received positive responses from the people living in the pilot area, i.e. that they could recognize the input that they had delivered during the participation process, and they confirmed that their ideas and solutions were incorporated in the final strategy. The strategy was formally adopted by the Provincial People's Committee in December 2013.

Notes

1 This book chapter is an extended and adjusted version building upon parts of B. Ottow, P. Huntjens and R. Lasage (2012) Participatief waterbeheer in Vietnam, *Water Governance*, 2(5): 32–37.
2 After the initiation phase, an extensive stakeholder assessment was carried out.
3 Long An Department of Science and Technology (DOST), Department of Agriculture and Rural Development (DARD), Department of Natural Resources and Environment (DONRE).
4 Members of the External Advisory Committee: 1) Prof. Dr. Stefan Kuks, Dyke Reeve of Water Board Regge en Dinkel and Professor in Water Governance at TU Twente; 2) Dr. Koos Neefjes, United Nations Development Programme (UNDP), Hanoi, Vietnam; 3) Dr. Bach Tan Sinh, Ministry of Science and Technology, Hanoi, Coordinator Asian Cities Climate Change Resilience Network; 4) Dr. Le Anh Tuan, Can Tho University, Coordinator MekongNet, Research Institute for Climate Change; 5) Prof. Ho Long Phi, Director of Centre for Water Management and Climate Change (WACC), Ho Chi Minh City; 5) Martijn van der Groep, Chief Technical Advisor, Mekong Delta Plan.

References

Alkan Olsson, J. and L. Andersson (2006) Models as a tool in water management, in A. Jöborn and I. Danielsson (eds) *Talking about water*, 131–148, Gothenburg: VASTRA Gothenburg University.

Andersen, D.F. and G.P. Richardson (1997) Scripts for group model building, *System Dynamics Review*, 13(2): 107–129.

Buuren, A. van and J. Edelenbos (2004) Why is joint knowledge production such a problem? *Science and Public Policy*, 31(4): 289–299.

Cockerill, K., H.D. Passell and V.C. Tidwell (2006) Cooperative modeling: building bridges between science and the public, *Journal of the American Water Resources Association*, 42(2): 457–471.

Cockerill, K., V.C. Tidwell, H.D. Passell and L.A. Malczynski (2007) Cooperative modeling lessons for environmental management, *Environmental Practice*, 9(1): 28–41.

Costanza, R. and M. Ruth (1998) Using dynamic modeling to scope environmental problems and build consensus, *Environmental Management*, 22(2): 183–195.

Costanza, R., F.H. Sklar and M.L. White (1990) Modeling coastal landscape dynamics, *BioScience*, 40(2): 91–107.

Dore, J. (2007) Multi-stakeholder platforms (MSPS): unfulfilled potential, in L. Lebel, J. Dore, R. Daniel and Y. Koma (eds) *Democratizing water governance in the Mekong region*, 197–226, Chiang Mai: Mekong Press.

Duc, N.H. and H.N. Minh (2008) *Vietnam: the effect of grassroots democratic regulations on commune government performance and its practical implications*, Ho Chi Minh National Political Administrative Academy, EADN Working Paper no. 35.2008, Ho Chi Minh City: Institute of Political Science.

Duong, M.N. (2004) *Grassroots democracy in Vietnamese communes*, research paper for the Centre for Democratic Institutions, Research School of Social Sciences, Canberra: The Australian National University.

Exter, K. den (2004) Integrating environmental science and management: the role of system dynamics modelling, PhD thesis, Southern Cross University, Australia.

Hare, M. (2003) *A guide to group model building: how to help stakeholders participate in building and discussing models in order to improve understanding of resource management*, Seecon Report # Seecon03/2003, Osnabrück: Seecon Deutschland GmbH.

Hare, M.P., O. Barreteau and M.B. Beck (2006) Methods for stakeholder participation in water management, in C. Giupponi, A.J. Jakeman, D. Karssenberg and M.P. Hare (eds) *Sustainable management of water resources: an integrated approach*, 177–225, Chichester: Edward Elgar.

Hart, A. (1986) *Knowledge acquisition for expert system design*, London: Kogan Page.

Hodgson, A.M. (1992) Hexagons for systems thinking, *European Journal of Operational Research*, 59(1): 220–230.

Hofstede, G., G.J. Hofstede and M. Minkov (2010) *Cultures and organizations: software of the mind: intercultural cooperation and its importance for survival*, 3rd edn, New York: McGraw-Hill.

Hostovsky, C. and V. Maclaren (2005) The role of public consultation in Vietnamese waste and other EIAs, paper presented at the First International Conference on Integrated Solid Waste Management in Southeast Asia, July 5–6, Siem Reap, Cambodia.

Huitema, D., M. van de Kerkhof, L. Bos-Gorter and E. Ovaa (2009) Public participation in water management: an analysis of innovative approaches from the Netherlands, in H. Folmer and S. Reinhard (eds) *Water problems and policies in the Netherlands*, 225–248, Washington, DC: Resources for the Future Press.

Huntjens, P. (2011) *Water management and water governance in a changing climate: experiences and insights on climate change adaptation in Europe, Africa, Asia, and Australia*, Delft: Eburon Academic Publishers.

Huntjens, P. and J. Bossert (2010) *Water governance cards: a water governance assessment pilot in Vietnam*. unpublished research report, Water Partner Foundation.

Huntjens, P., C. Pahl-Wostl, Z. Flachner, S. Neto, R. Koskova, M. Schlueter, I. NabideKiti and C. Dickens (2011) Adaptive water management and policy learning in a changing climate: a formal comparative analysis of eight water management regimes in Europe, Asia, and Africa, *Environmental Policy and Governance*, 21(3): 145–163.

Huntjens, P., L. Lebel, C. Pahl-Wostl, R. Schulze, J. Camkin and N. Kranz (2012) Institutional design propositions for the governance of adaptation to climate change in the water sector, *Global Environmental Change*, 22(1): 67–81.

Korff, Y. von (2005) *Towards an evaluation method for public participation processes in AquaStress and NeWater: a proposal for both projects*, AquaStress and NeWater internal working document, Sixth EU Framework Programme, Montpellier: Cemagref.

Lebel, L., S. Neto, P. Huntjens and J. Camkin (2011) Critical reflections on multi-stakeholder dialogues on water: experiences in the Netherlands, Australia, Mekong Region, and Portugal, in P. Huntjens (ed.) *Water management and water governance in a changing climate: experiences and insights on climate change adaptation in Europe, Africa, Asia and Australia*, 211–230, Delft: Eburon Academic Publishers.

Mekong Economics (2006) *Measuring grassroots democracy in Vietnam*, Hanoi: Mekong Economics/Embassy of Finland.

Pahl-Wostl, C. (2007) Transition towards adaptive management of water facing climate and global change, *Water Resources Management*, 21(1): 49–62.

Palmer, R. (1999) Modeling water resources opportunities, challenges, and trade-offs: the use of shared vision modeling for negotiation and conflict resolution, paper presented at ASCE 26th Annual Conference on Water Resources Planning and Management, June 1999, Tempe, AZ.

Reed, M.S. (2008) Stakeholder participation for environmental management: a literature review, *Biological Conservation*, 141(10): 2417–2431.

Rouwette, E.A.J.A., J.A.M. Vennix and C.M. Thijssen (2000) Group model building: a decision room approach, *Simulation and Gaming*, 31(3): 359–379.

Rowe, G. and L.J. Frewer (2000) Public participation methods: a framework for evaluation, *Science, Technology & Human Values*, 25(1): 3–29.

Stave, K.A. (2002) Using system dynamics to improve public participation in environmental decisions, *System Dynamics Review*, 18(2): 139–167.

Vennix, J.A.M. (1999) Group model-building: tackling messy problems, *System Dynamics Review*, 15(4): 379–401.

Vries, F.W.T.P. (2006) Learning alliances for the broad implementation of an integrated approach to multiple sources, multiple uses and multiple users of water, *Water Resources Management*, 21: 79–95.

Zagonel, A.A. and J. Rohrbaugh (2007) Using group model building to inform public policy making and implementation, in H. Qudrat-Ullah, J.M. Spector and P. Davidson (eds) *Complex decision making: theory and practice*, 113–138, New York: Springer.

5 Understanding institutionalized ways of knowing climate risks

Reflections on action research for participatory knowledge production

Daan Boezeman, Martinus Vink and Pieter Leroy

Introduction

6 January 2012, a dyke near Woltersum, northern Netherlands, is about to collapse. Water and sand flow through the dyke, threatening its stability. A combination of intensive rainfall and a north-western storm on the Wadden Sea is hampering the discharge of water. Water in the *boezem* systems, which are interconnected networks of water courses and lakes to store excess polder water, reaches peak levels. About a thousand inhabitants are evacuated. The January 2012 event fits in a series of recent high waters in this part of the Netherlands: 1993, 1995, and 1998. In 1998, the situation was serious: near the city of Groningen water flowed over dykes, a dyke near Winschoten almost collapsed, and three polders were inundated to prevent the flooding of built-up areas. The inundation damaged Gasunie's gas infrastructure and led to huge damage claims.

The events of the 1990s provoked a policy response called HOWA, High Water, a project initiated in 1999, comprising policies costing around €200 million. Despite the sense of urgency and the consensual traditions of Dutch water management, HOWA's policy strategy led to controversy and resistance, in which property rights and the allocation of costs and benefits, as well as the decision-making process itself, were quintessential. A decade later, the provinces of Groningen and Drenthe announced a follow-up project, later called Dry Feet 2050 (DV2050), meant to:

> [S]tudy whether and what additional policies are needed against water nuisance coming from the boezem to live up to the safety norms ... desired in 2025. Those policies need to contribute in 2050 and 2100 to the safety, and must therefore be sustainable, 'no regret policies'. ... besides a study on climate adaptation policies, we want to actualize safety norms based on the latest spatial development [and] soil subsidence due to natural gas drilling.
>
> (Letter to Provincial Council, Groningen,
> 21 January 2011, our translation)

The DV2050 study area includes the catchments of the Noorderzijlvest and Hunze & Aa's water boards, both located partly in the province of Groningen

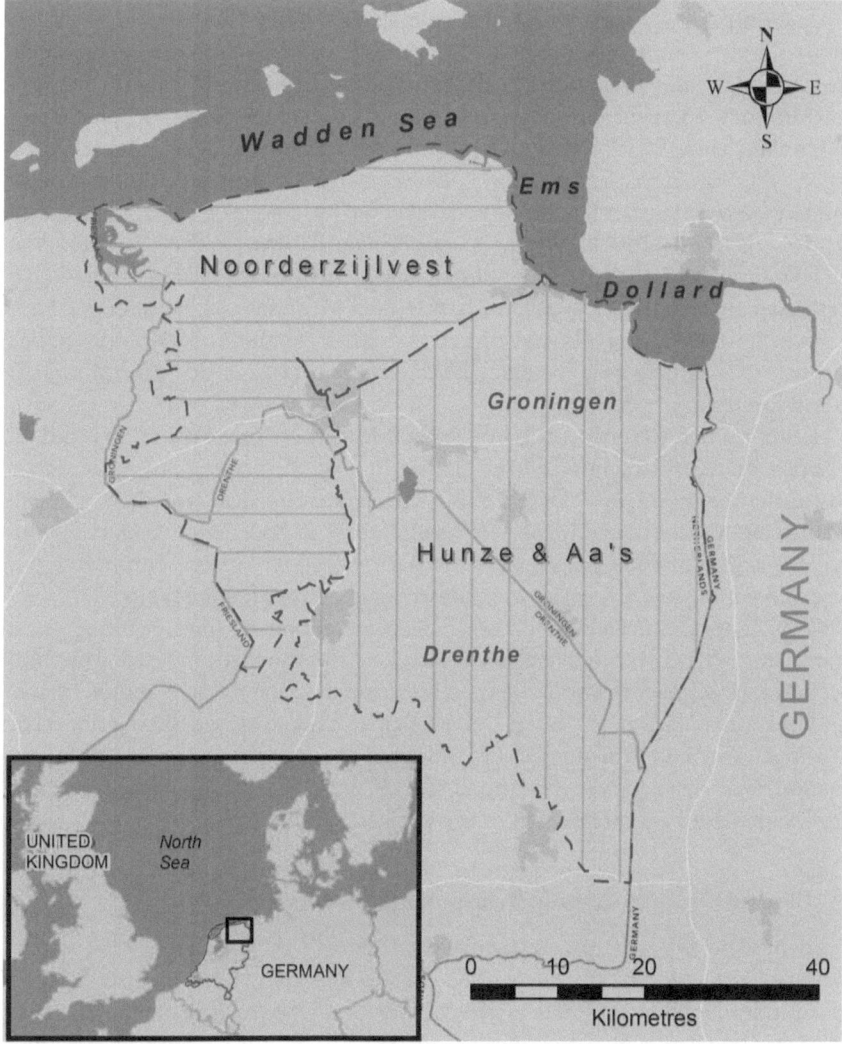

Figure 5.1 Catchment areas of Noorderzijlvest and Hunze & Aa's

and partly in Drenthe (Figure 5.1). Although the overall goal is in line with previous projects and initiatives, two things stand out: beyond actualizing norms to ongoing changes, climate change was the new substantive element, and, because of the earlier controversies, the Dry Feet 2050 process was to take a more participatory approach. Mainly for the latter reason, the province of Groningen and the Noorderzijlvest and Hunze & Aa's water boards invited us as researchers to take part in DV2050 in order to learn and reflect on stakeholder participation. The former reason made DV 2050 an intriguing case of (participatory) knowledge production and policymaking for climate adaptation.

The intriguing character of the case derives from the following: in the Netherlands, water management in general, and water risk governance in particular, are among the most established regimes, dominated by well institutionalized agencies, and largely formalized competencies and procedures (Bourblanc *et al.*, 2013; Kuks, 2009). In this context in particular, it is intriguing to observe how climate adaptation policies emerge. Climate adaptation is not a field in its own right; rather, the issue is taken up by a range of pre-existing policy fields, a phenomenon known as *mainstreaming*. However, the understanding of how pre-existing institutions – in our case, institutions dealing with almost perpetual flooding risks – deal with this new phenomenon in terms of both knowledge production (Boezeman *et al.*, 2013; Jasanoff, 2004) and policy responses (Inderberg and Eikeland, 2009; Rayner *et al.*, 2005; Vink *et al.*, 2013) is still limited.

Being invited to support the DV2050 project, therefore, gave us a privileged position to elucidate our central question: how do pre-existing governance institutions respond to new demands – in our case to deal with both climate adaptation and participation as new requirements in knowledge production and policymaking? In addition, we were curious to see in what sense our own action research could have an impact on mainstreaming either requirement.

This chapter proceeds as follows. The next section outlines our theoretical frame and reflects on the role of the researchers and the methods used. The third section introduces the region. We then present our action research interventions, and the organization and scope of participatory knowledge production that resulted. The penultimate section interprets this course of action from our institutional perspective. The final section reflects on the advantages and disadvantages of action research in this project.

Action research with an institutional perspective

Studying the DV2050 project while participating in it as action researchers required a double theoretical perspective: we needed (1) an understanding of institutions and how these deal with (calls for) innovative action and (2) insights into the possible roles of action researchers. Insights from both perspectives informed us in the gradual development of our role and stance in the project, as we now report.

Institutions and their (in)ability for innovative action

Studying the collective action initiated by the DV2050 project group, first requires a theory of action. Institutional perspectives on action challenge rationalist approaches in stressing that actors interpret events in structures with which they are socialized, reproducing regulative, moral, and cognitive boundaries in behaving in appropriate manners (March and Olson, 2006; Scott, 2008). Institutions, like those well established in the Dutch water management domain, are regarded here as 'those social patterns that, when chronically reproduced, owe their survival to relatively self-activating social processes' (Jepperson, 1991: 145). Courses of collective action have materialized in roles and competencies for actors, formal and informal rules, organizations, procedures to be followed to arrive at credible and legitimate answers, and so forth. These constellations show path dependency, have grown historically, and structure subsequent action. Non-conformity to routinized practices is considered to be costly in various ways: it increases economic risks, decreases the predictability of the process, requires more thought, or reduces legitimacy or access to those resources that are attached to perceptions of legitimacy (Phillips *et al.*, 2004). Hence, institutions constrain and enable action, including the (innovative) action that DV2050 demands. The consequent empirical question, therefore, is the extent to which and by what mechanisms the institutions involved leave room for that innovation, e.g. for a more participatory approach to take place.

According to Jasanoff (2004: 40), knowledge production – in our case, the development of a shared understanding of the risks of climate change – also evolves along institutional patterns. Knowledge is not so much stored in the mind but rather tacitly comprehended in carriers such as symbols, objects, language, and classifications that tend to be firmly rooted in all sorts of well-established institutional practices and processes, yet in turn can be mobilized in new situations. Stated differently, 'cognitive abilities do not reside in "you" but are distributed throughout the formatted setting' (Latour, 2005: 211). New accounts are constructed with what is already there. Here again, then, the empirical question is the extent to which and by which mechanisms institutions are open to learning (Argyris and Schön, 1989).

Building on Scott's (2008) conceptualization of the regulative, normative, and cognitive aspects of institutions, we focus on three mechanisms in which the pre-existing institutions constrain and enable the response to the new requirements of increased participation and climate adaptation:

- rules specifying decision competencies, tasks, and roles of the different actors involved;
- normative arguments articulated to legitimize certain courses of action; and
- procedures, guidelines, and standardizations invoked to facilitate the assessment of water risks.

Roles and practices of researchers

Participating in action research requires a critical reflection on the role that the researchers want to, and can, take. Research on scientific experts (Pielke, 2007) or knowledge brokers (Turnhout *et al.*, 2013) distinguishes among a multitude of roles for researchers when they are engaging in the political process. As a simple heuristic device, Turnhout *et al.* (2013) plot ideal type roles on a continuum (Figure 5.2). In the most extreme form of action research, the boundary between science and policy, as well as those categories as distinct entities themselves, would dissolve. Still, individual researchers may take different roles or switch between them.

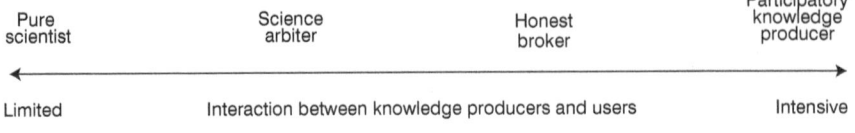

Figure 5.2 Roles of researchers and intensity of interaction (Turnhout *et al.*, 2013)

Turnhout *et al.* (2013) observe different practices and repertoires of policy analysts and researchers at the boundary between science and policy. In a 'supplying' repertoire, a researcher predominantly brings in science or other experts to the project. In the 'bridging' repertoire, the researcher tries to mediate: helping to articulate questions and translating science into meaningful knowledge. When 'facilitating', scientific knowledge is no longer central, but rather helps to develop the user's own understanding of the situation at hand. Below, we elaborate on the role that we envisaged and the methods we used to pursue that, and reflect upon the impact of our action in the concluding section.

Research strategy, methods used, and roles taken

When we were invited by the water boards and provinces to take part in DV2050 in order to learn and reflect on stakeholder participation, we were able to draw on a body of knowledge on participation, but lacked an in-depth understanding of the idiosyncrasies of water management practices in this region. Tacit knowledge about how practitioners deal with regulation in practice, latent power struggles, or past participatory experiences in the region were unknown to us when we started. Therefore, we followed an action research strategy that was open, iterative, and aimed at facilitating a learning experience on participation. We never advocated a particular form of participation as best, nor was our approach aimed at empowering particular stakeholders in the region. We had an observant and reflective role and used elements of *action inquiry* (Argyris and Schön, 1989) and *learning evaluation* (Huntjens *et al.*, Chapter 2, this volume). Over an intensive two and a half years, we went through multiple cycles of observation, conclusion, and (re)action to enhance reflection as the project developed.

We fully entered the scene while organizing two workshops (summer and autumn, 2011). These workshops dealt with the participatory approach for which DV2050 would opt. We discuss the workshop substance in the fourth section, but we highlight them here as emblematic junctures in the role that we as researchers took: bridging and facilitating. Rather than promoting a particular participatory approach, we used the stakeholder participation guide (Hage *et al.*, 2010) which one of the authors of this chapter co-authored for the Netherlands Environmental Assessment Agency. The guide does not function as a standardized cookbook, but aims to enable a critical reflection by those working with it on the motives (why?), scope (what?), intensity (how much?), intended participants (with whom?), and form (how?) of their participation project. In our interactions with project members, we repeatedly discussed these questions, asking them for clarification on their viewpoints, inviting them to reflect on the advantages and pitfalls of every alternative approach on these and subsequent questions, and leaving them to decide. Only at their explicit request did we provide the project members with insights from additional literature. Hence, our role resembles the *honest broker* ideal type, as we actively engaged with users, yet maintained a distinction between researcher and policymaker, each with its own role and responsibilities.

We used action research tools in combination with regular methods of interpretative policy analysis (Yanow and Schwartz-Shea, 2006). A content analysis of documents of projects prior and related to DV2050 strengthened our understanding of the developments that led to the start-up of the DV2050 project, and enabled us to trace innovations and deviations in DV2050 compared with past processes. These were regarded as indicators for change, innovation, and learning. We studied project documents and internal correspondence on DV2050 itself. Next, we engaged in participant observations of around 20 project meetings of DV2050 and five stakeholder meetings. All these were audio recorded. Throughout the whole process, we used a reflection diary with preliminary observations and interpretations. Not only did we compare these one- or two-page documents with our later interpretations, but these notes were also part of the frequent face-to-face and email conversations with the eight project members over the course of the project. Furthermore, we interviewed members of previous projects, as well as around 20 stakeholders (shorter interviews) during participation meetings. This was done in order to reconstruct these actors' interpretations of the unfolding process. We also surveyed stakeholders jointly with Noorderzijlvest to gauge opinions on the project. Finally, we had rounds of reflection interviews with the project members. These semi-structured interviews discussed motives for courses of action. We 'member checked' our preliminary analyses as well as scholarly texts to let project members comment on them. We also shared preliminary analyses with a peer group of action researchers on adaptation governance.

Introducing the region, introducing the issues

The DV2050 project deals with the catchment areas of the Noorderzijlvest and Hunze & Aa's water boards, respectively 144,000 ha and 207,000 ha in size, and with 375,000 and 420,000 inhabitants. The catchment area is divided in several *boezem* systems. These *boezems* receive water from the adjacent polders and brooks in the south, temporarily store that water, and transport it to either the Wadden Sea, Ems, or Dollard in the north. A large part of the area is below sea level. Therefore, water is pumped from the polders into the *boezem* to maintain water quantity to meet agriculture's requirements. *Boezem* water is either pumped or discharged into the sea at low tide.

From a flooding perspective, levees and storage areas are crucial, all the more so as many of these levees are designated as regional water barriers. Noorderzijlvest has about 400 km and Hunze & Aa's 500 km of these barriers. Flooding is said to occur mainly when a period of increased precipitation coincides with north-western storms, temporarily hindering the discharge of excess water. Climate change is thought to intensify this issue, as Dutch regional climate scenarios suggest accelerating sea-level rise and increased winter precipitation. This is further complicated by other developments mentioned in the DV2050 starting paper: soil subsidence due to mining for natural gas, settling of the ground, and peat oxidation.

Action research in the Dry Feet 2050 project

This section first sketches our interventions in relation to the design of the stakeholder participation in DV2050. Second, we present how the project group responded to the double demand mentioned earlier: increasing participation and inserting climate adaptation. Hence, this second part describes the actual results of the action research process to design a participatory process.

Action research on a stakeholder participation strategy

Developing shared knowledge on stakeholder participation

Early in 2011, we first contacted DV2050. At that time, policymakers at the Noorderzijlvest water board were starting up a joint project with the provinces of Groningen and Drenthe and the Hunze & Aa's water board to develop policies for the future of the water system in the face of soil subsidence and climate change. While doing so, the DV2050 project members formulated a demand for advice and reflection on how to organize effective stakeholder participation. Participation was said to be a rather new element – contradicting the conclusions of several scholars (Van Buuren et al., 2012; Van den Brink, 2009) that stakeholder participation was a general trend in Dutch water management.

At the project members' request and informed by the 'why' question in the stakeholder participation guide referred to above, we presented literature on participation and led a subsequent discussion to develop a shared understanding of the motives for organizing participation. The project members articulated diverging motives for engaging stakeholders in DV2050. First, some stated that participation would lead to greater acceptance for the policies resulting from the project. Early participation would therefore save time in later stages. From the preceding high-water assessment project, HOWA (1999–2002), they learned that the lack of participation had provoked a lot of resistance and controversy. This was to be avoided by bringing powerful regional actors on board, e.g. the nature conservation and farmers' organizations, to commit them to the outcomes of DV2050. Second, stakeholders were said to potentially have valuable information and/or innovative solutions. Third, some project members articulated normative arguments: involving stakeholders was simply a matter of good governance. In brief, the three classical motives for participation (instrumental, qualitative, normative) were evoked, albeit with a clear emphasis on the former.

At that time, however, we already observed tensions between the water boards and the provinces on their respective competencies in DV2050. According to the Noorderzijlvest water board project members, the latter 'can and wants to have space to push integrated water management in the project, whereas the province thinks of Noorderzijlvest as the supplier of technical knowledge' (meeting, 15 March 2011). The province was well aware of this divergence with Noorderzijlvest, yet seeing its causes in '... the changing role of the water board Noorderzijlvest with which they are struggling internally' (meeting, 29 June 2011). We return to this divergence on ascribed roles below.

Delineating the 'stakeholders'

In July 2011, we organized a workshop to respond to the project group's query about 'who to invite.' In this period, there were high tensions in the project group between the water boards and the provinces. In the workshop, we aimed to develop a shared view on at least the *possible* actors to be engaged in the project, while leaving open the opportunity *not* to invite stakeholders directly, yet to design the process in such a way that interested people could easily join during open ateliers and similar working methods.

The workshop was based on a simple social map. In two groups, the project members made a list of all possible actors they could imagine. They then plotted all actors on the map depending on the perceived interest that particular actor had in the DV2050 project. Interestingly, the two groups had roughly the same maps and could rather constructively explain their differences. The results, therefore, were merged in one single map. We considered every actor, resulting in a short profile for each of them. The short report we wrote was shared with the project group, but no fundamental amendments were made. The map itself was used to invite stakeholders for an initial meeting of the project in September 2011, during which the two professors involved gave presentations on participation for project members, stakeholders, and administrators, and discussed the issues with the audience. In retrospect, these presentations and discussions endorsed and legitimized not only our position as action researchers, but also the choices the group had made there.

Developing a stakeholder interaction scheme

In November 2011, we organized a second workshop, focusing on the 'how?' and 'how much?' of the participatory process. On the basis of various practical methods in the stakeholder guide and related literature, we compiled lists of options for forms of participation (ranging from newsletters, to hiring liaison officers and policy design ateliers). The objective that we co-developed in preparatory individual meetings with representatives of both the province and the water board was to design a shared 'process architecture' and to elaborate that in concrete elements.

This workshop, unlike the former, failed to reach its objectives. Various DV2050 members were very sceptical about the need for participation and about the risks of different routes. A shared design was not agreed upon, and the meeting was frustrating. In particular, the hitherto somewhat latent divergences between Noorderzijlvest and Groningen became very visible now: they differed on the variety of stakeholders to be involved, on their impact, and on a series of adjacent issues. Each party threatened the other that they would use various means of power at their disposal, e.g. by not contributing financially to parts of the project, up to withdrawing from the DV2050 project altogether.

This discussion was eventually settled in early 2012 by deciding on three different, but layered, forms of participation in DV2050 (explored further in the next section). What is important here is that this outcome reflected and reproduced the formalized competencies of the different government bodies involved: the body with the formal right to decide on the issue also claimed to have the right to decide on the process in which that decision was going to be made. In other words, although participation was hailed as an innovation, none of the parties would allow that participation to trigger any change in their respective competencies.

From that point onwards, we did not organize any more workshops, but provided comments and reflections upon documents and communication plans. Of course, we continued our observations on how the group dealt with the participation issue, and the reasons that were formulated for choosing a particular organizational form over another. We shared these interpretations – see below – in a series of reflection interviews with the project members, in late 2012 and early 2013.

Stakeholder participation in Dry Feet 2050

The organization of participation

The actual organization of the DV2050 stakeholder participation became fragmented in two ways. First, participation-by-invitation was organized differently in the various sub-parts of the project (Figure 5.3 sketches all sub-groups with their tasks, competencies, and different participatory strategies). The general DV2050 project group took a coordinating and communicative role with both internal and external stakeholders, the former being various parts of the involved governments, the latter being non-state stakeholders. On the one hand, it aimed to build the support of influential stakeholders by involving them actively; on the other hand, it coordinated the communication with all stakeholders via a bi-annual newsletter, a Twitter account, and an annual stakeholder participation day. The workgroup chaired by Noorderzijlvest called stakeholders to adopt an active role and a shared ownership. Their aim was to connect goals and ambitions of the water board and stakeholders in workshops on policies. The workgroup chaired by Hunze and Aa's did not organize much interaction with stakeholders.

Second, all sub-groups made a distinction between 'priority' and 'other' stakeholders when organizing the invitations. A stakeholder was labelled as priority when the organization either owned and managed land itself (e.g.

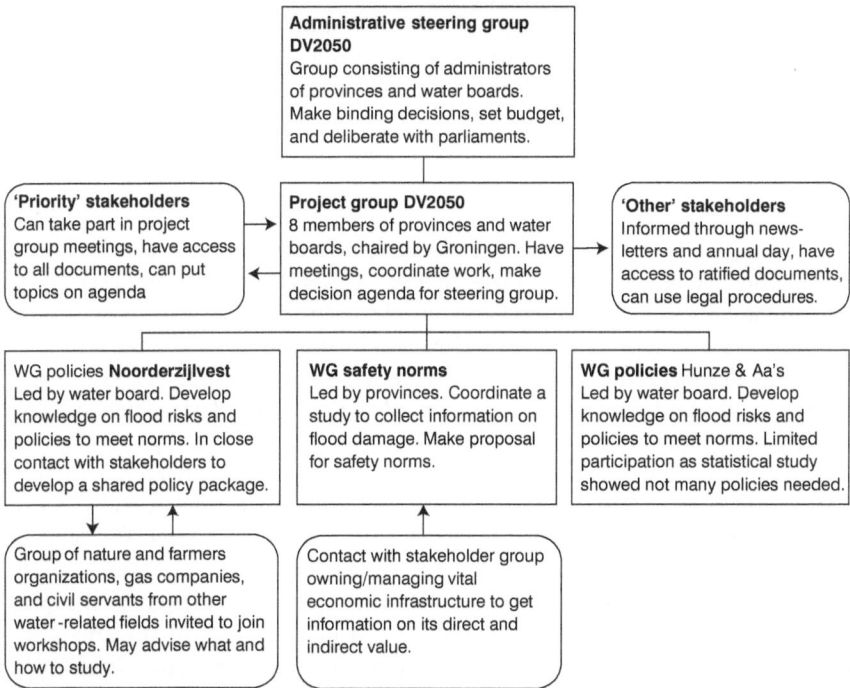

Figure 5.3 Overview of the DV2050 project specifying tasks, competencies, and stakeholder participation strategy of all sub-groups

agriculture, nature, or large industrial sites), represented a group of actors that had acted as opponents in the past and it was thought that they might do so again (farmers, municipalities, and environmental organizations), or had to contribute financially to policies (e.g. the committee on soil subsidence due to gas drilling). Most, but not all, of these actors were plotted in the centre of the stakeholder map, and generally project members knew their representatives personally from earlier projects. These stakeholders were invited and given many opportunities to directly influence DV2050: amend the agenda, provide input on the criteria on which policies are evaluated, and state which options they support or oppose. The other stakeholders were most of the other organizations that the group indicated on the periphery of the stakeholder map. These stakeholders received the newsletter and were invited to the annual participation day. Citizens were deemed to be represented via their municipalities, farmer organizations, or environmental groups.

The scope of participation

As to the 'on what?' (to participate) question, the scope was clearly demarcated as follows: what packages of policies are best taken to meet the safety norms, given the consequences of climate change and soil subsidence? Although

Noorderzijlvest's initial scope was far more ambitious, as echoed in the Water System Management 2050 proposal, the scope was eventually firmly narrowed to the above. Thus, the 'problem' was given, and stakeholders were expected to contribute to policies they thought acceptable, e.g. storing water longer in polders and nature areas, raising dykes, creating inundation polders, building a large discharge pump, and so on. One issue was clearly beyond the project's scope: the safety norms. The province would decide upon them solely, without allowing any participation in relation to norm-setting (see lower middle part of Figure 5.3). Other likely effects of climate change, such as drought, water quality, or short summer peak precipitation floods, were considered beyond the scope of the project – even though these issues were raised by stakeholders.

In conclusion: pre-existing institutionalized patterns of competencies and pre-existing patterns of interaction largely delineated the room of manoeuvre for any participatory innovation, with regard to both its processing and its substantive scope. As a consequence, DV2050's participatory aspects differ only slightly from previous projects. Stakeholders are now involved earlier in the process, and they may contribute their visions and information. However, the priority stakeholders – farmers and nature conservationists – have always been strongly involved and influential. Therefore, we observe a reproduction of the pre-existing neo-corporatist pattern in Dutch water management (Bourblanc *et al.*, 2013), in which experts strongly influence the space for participation (Halffman, 2009). The delineation of relevant stakeholders or the substantive scope of water management is not easy to change (Van Assche *et al.*, 2011). Participation, therefore, can be considered an extra element to this project, without it leading to any reconfiguring of knowledge production or decision making (cf. Van Buuren *et al.*, 2012).

Understanding the policy process through the institutional lens

To elucidate why both the processing and the substantive scope of the project were hard to change, we now turn to the highly institutionalized and codified institution of regional water safety management. We refer to Scott's (2008) analytical distinction between the regulative, the normative, and the cognitive aspects of institutions and institutional change.

The regulative aspect: formalization and controversy in regional flood risk management

Until recently, there were no flood norms for the regional barriers, in particular the *boezem* levees. In the Fourth Memorandum on the Water System (1998), the minister decided to oblige the provinces and water boards to develop testable safety norms for these regional barriers and to periodically reassess whether those norms are met. This system was formalized in the 2009 Water Act. Because of the complexity of understanding flood risks, the system is constructed around the simple and quantifiable notion of the threshold probability: the probability that the water level in the *boezem* might exceed the design high-water level that a barrier must safely contain. This is the safety norm. Two activities are crucial and mandatory. One, every possible state of nature has to be translated into water levels in the *boezem* expressed in terms of 'a likelihood of occurrence'. To make these assessments possible, standardized procedures were developed on how this should be done (to which we return later). Two, the threshold probability has to be determined. The so-called IPO-method, named after the umbrella interprovincial organization, serves as a standardized assessment guideline with damage categories to set standards based on the economic value behind the levee: the higher the economic value damaged by flooding, the higher the norm, thus, the lower the probability should be. If a water level's probability is higher than the norm, either policies have to be implemented or the norm has to be lowered.

In addition to this standardization of safety measurement, tasks and roles are formalized. The National Government Accords on Water (2003, 2008, and 2011) specify agreements on how to organize the risk-assessment process and how to keep the regional water system in good order in 2015 and beyond. The province is the authority that has to specify the safety norms for the *boezem* systems within its territory. As the manager of the catchment area, the water board is responsible for assessing the flood risks of the *boezem* and for developing policies to meet these norms, supervised by the province. The water board has to fund those policies with its water board taxes.

These formalized competencies create tensions between provinces and water boards. The former can impose higher safety levels than the latter may regard as feasible and affordable. Furthermore, the province may treat the boards as an auxiliary technical agency for its decisions on water management. The water board, on the other hand, strives to develop its integrated water policies as an independent government body. The respective roles and tasks of these two government bodies have been redefined in the 2011 Water Accord, yet both actors still struggle over the exact interpretation thereof. A similar struggle had partly overshadowed the DV2020 predecessor project, HOWA, in 2001. In brief: the province had set a uniform safety norm for its entire territory that was higher than the IPO method suggested, and the water board felt that it had to meet higher targets than it deemed desirable itself. This caused controversy, irritation, higher costs, and an erosion of trust. During the project meetings, project members repeatedly referred to HOWA, as they experienced the DV2050 process as an unpleasant repetition of previous events.

In this respect, the need for DV2050 resulted partly from the periodical obligation to test the regional safety norms. The continued power play over competencies between the province and the water board overshadowed the project, including its initially envisaged participation. DV2050 turned out to be a fragmented rather than a genuinely joint project. And regarding participation and scope, the province firmly preserved its competencies on setting the norms, the water boards on deciding on policies. Therefore, any stakeholder participation that could threaten these respective positions had to be prevented. Hence, DV2050 could not have any innovative potential.

The normative aspect: the status quo as a legitimate course of action

The normative pillar emphasizes the role of values and norms in creating expectations and obligations, such as role expectations. Although there was clearly a power struggle between the province and water board, in general the legitimacy of DV2050's way of operating was not seriously challenged. Regarding the substance of the project, the central value was fixed: the degree of flood protection provided by the government was to be increased, climate change providing an extra argument. Also, the idea that those areas with more accumulated capital (industrial estates or dense residential areas) would be better protected than other areas was considered more appropriate than providing equal protection to all. These values were not disputed by stakeholders, or by political parties when the provincial parliament dealt with the DV2050 project.

Stakeholders did not dispute the legitimacy of the governance process. Together with Noorderzijlvest, we conducted a survey in the workshops among all invited priority stakeholders (23 participants, response rate 83 per cent). The results suggested that there was no need to change the respective governance roles of stakeholders and governments. Stakeholders did not desire a too influential position with responsibility for the joint solution: 0 per cent preferred just information, 17 per cent consultation, 61 per cent advice, 22 per cent co-producing, and 0 per cent self-steering. Interviews (Noordzijlvest, 30 October 2013) suggest that, even though Noorderzijlvest's sub-project offered more space for participation, historical role expectations of what every actor could and should appropriately do proved very hard to change. Many of the appropriate roles and ways of operating are also formalized in laws and standard procedures. Without serious challenges to its legitimacy, the current path of water management was reproducing its status quo.

The cognitive aspect: the instruments enabling flood risk assessments

To enable the aforementioned quantitative assessment of flood risks, a range of tools and guidelines were developed. Harmonization projects of IPO and STOWA (the water boards' knowledge organization) seek to standardize the methods for norm setting and testing. These risk-assessment procedures greatly reduce the complexity of the project group's tasks, by simplifying its natural and organizational environment. For instance, it enables the division of labour between a group studying how high the flood risks need to be according to the 1999 IPO methodology and a group studying whether the water levels corresponding to those norms will be met with the proposed measures. The pre-set assessment instruments reduce these complex tasks to rather simple applications of standardized methods.

The same reduction of complexity applies to the role of the hydrological and hydraulic models with which the water boards work. The shared beliefs in the soundness of that approach are illustrated by the project plan of the water board stating that:

> [t]here is no alternative to calculation models. An event occurring 1:100 years cannot be grasped in the practical experience of field staff [and] practical knowledge is insufficient to make good policies.
>
> (Noorderzijlvest, 2012: 4, our translation)

The models concerned are loaded with information on (the parameters of) canals, retention areas, sluices and pumps, and so on. In earlier studies, substantial time and financial resources were invested to refine and verify these models. These models enable the expression of every possible state of nature in water levels in the *boezem* in terms of likelihood of occurrence.

The group standardized probabilistic methods: individual stochastic variables, such as initial groundwater levels, precipitation, and storm tides, each have a certain probability and associated value. Combinations of these stochastic variables yield hydrological forcing with frequency of occurrence in the *boezem*, for which their SOBEK model then calculates water levels in different locations in the water system. The data used are supplied by a web of local measuring stations and standard extreme precipitation statistics certified by the Royal Netherlands Meteorological Institute and HKV Consultants in 2004.

Taken together, the institutionalized risk assessment procedures enable location-specific knowledge production on quantitative threshold probabilities within a limited time frame. The assessment of climate impacts is concentrated mainly around this practice within this quantitative water safety problem framing. Hence, what is or is not the problem, or what can or cannot be calculated as effects of climate change, is constrained by these modelling procedures. We saw earlier that stakeholders raised other climate effects as important for the assessment – because they might have an effect on dyke stability and hence flooding – but were considered to be beyond the scope of the project, as these effects were not considered by experts as influencing the relevant water levels. In theory, it is possible to change the given problem framing, but it needs an enormous amount of work to do so in a hydrologically legitimate way. As Rayner (2003: 168) observed, in a political culture of science-based expertise, science sets the stage for participation and frames the issue as hard-edged, making other perceptions appear malleable.

Our first research question was how pre-existing governance institutions respond to new demands. Despite the power struggle between Groningen and Noorderzijlvest, we conclude that the absence of serious legitimacy disputes on the one hand, and a strong alignment between the regulative, normative, and cognitive aspects of the governance institution on the other, yielded a reproduction of the status quo. The pre-existing governance institutions structured the participatory knowledge production on adaptation in DV2050, but those new demands did not erode the stability of the institutions themselves.

Discussing the potential of action research

We now turn to our second research questions and discuss the extent to which action research led to an improved understanding of, and had an impact on, the DV2050 process.

Action research and its potential for understanding DV2050

For us as researchers, setting up this endeavour as an action research project created a range of opportunities. First, it allowed us to actually do observations within an organization translating climate change in the direct context of decision making. Not only was learning on 'increased participation' our entrance ticket for 'being there' (Yanow and Schwartz-Shea, 2006), our action research through joint efforts such as the workshops or the stakeholder survey provided interesting research results and were only possible because of the intensive interaction. Information was shared with us in a way that would possibly not be the case with classical retrospective interviewing. Second, these close interactions and the mutual trust to which they contributed helped to make tacit knowledge explicit about appropriate modes of conduct for the different actors in regional water management. They provided the opportunity to observe how formalized rules specifying competencies were brought into the process by the actors involved. We thus saw the actors themselves re-actualize the institutional constraints that they said were relevant to them. And we saw and heard these actors draw analogies to past procedures that proved either success or failure. In that sense, Lewin's conclusion – one of the best ways to understand the world is to try to change it – proved very valid here. Third, the frequent and continued interactions offered the opportunity to test whether our conclusions were missing important elements, for example on the power struggle over competencies or the stubbornness of role expectations. 'Normal' case study research seldom offers these opportunities.

Engaging in an action research project also has potential drawbacks. The time investment in action research is substantial. In our case, it meant a continued effort over a long period with many intensive episodes. Although it was worth the effort in this particular case, investing in such a detailed understanding is not necessary for every research problem. Also, action research, despite its huge investments, means accepting the uncertainty about what the jointly agreed research questions will be.

Action research and its impact on DV2050

We were invited to take part in DV2050 to help to learn about increased participation. Since no substantial change towards participation was realized, one could conclude that our action research intervention failed, i.e. has had a limited impact on the DV2050 process. This, however, seems a premature and unfair conclusion. From what we have observed ourselves, and from reflections by people involved, we have had an impact on various levels.

First, on an instrumental level, some of the jointly produced knowledge, such as the stakeholder map and the considerations that were formulated on the basis of it, not only circulated in various policy documents but was also used to organize the invitations. In a similar fashion, several elements of the second workshop were implemented in the process design. Second, and maybe more important, are aspects of strategic impact and use of our presence and action. We as researchers, in an involved yet distanced role, had a legitimizing effect. For the project group, getting confirmation that it was on the right track was important; and this group strategically used us to legitimize the project vis-à-vis the stakeholders: 'showing we had support of researchers was a sign to stakeholders during the kick-off day that we were taking that matter seriously' (interview, Groningen 7 March 2013).

While we positioned ourselves in a bridging role, bridging between academic knowledge and policy practitioners, we unintentionally also functioned as an independent academic arbiter 'at a distance providing a second opinion to the advice of hired consultancy firms' (interview, Drenthe, 7 March 2013). Some respondents even pointed at a stabilizing function that we – unintentionally – had within the project group, in particular at difficult junctures. Our rather 'neutral' role at some distance assuaged the thorny power interest struggle.

A further goal of the project was to reflect on, and learn about, effective participation. Several interviews suggest that reflection was highly appreciated. Learning was restricted mainly to first order learning: the pros and cons of different strategies and instruments to be used for participation, whereas second order learning (Argyris and Schön, 1989) on values, assumptions, or theories underlying participation was limited. In this case of powerful competing interests and the strongly institutionalized governance patterns, there were few opportunities to experiment with other stakeholder approaches.

In retrospect, given the strongly institutionalized context in which this action research project evolved, we believe that only two roles were sustainable: either take a consultancy stance, advising on a certain approach to participation – a stance that we deliberately rejected – thereby in fact supporting one of the visions advocated by the parties involved, or maintain a rather 'neutral' position from which we could facilitate deliberation and debate among these parties. That we opted for the latter does not mean our influence was neutral in its effects: some interviews suggested that our stakeholder map actually justified the idea that one can indeed distinguish stakeholders as 'priority' or 'secondary,' thereby in turn reconfirming existing power relations. Again: the best way to understand the world is to try to change it.

References

Argyris, C. and D. Schön (1989) Participatory action research and action science compared: a commentary, *American Behavioral Scientist*, 32(5): 612–623.

Assche, K. van, M. Duineveld, R. Beunen and P. Teampău (2011) Delineating locals: transformations of knowledge/power and the governance of the Danube delta, *Journal of Environmental Policy & Planning*, 13(1): 1–21.

Boezeman, D., M. Vink and P. Leroy (2013) The Dutch Delta Committee as a boundary organisation, *Environmental Science & Policy*, 27: 162–171.

Bourblanc, M., A. Crabbé, D. Liefferink and M. Wiering (2013) The marathon of the hare and the tortoise: implementing the EU Water Framework Directive, *Journal of Environmental Planning and Management*, 56(10): 1449–1467.

Brink, M. van den (2009) *Rijkswaterstaat on the horns of a dilemma*, Delft: Eburon.

Buuren, A. van, E.H. Klijn and J. Edelenbos (2012) Democratic legitimacy of new forms of water management in the Netherlands, *International Journal of Water Resources Development*, 28(4): 629–645.

Hage, M., P. Leroy and A.C. Petersen (2010) Stakeholder participation in environmental knowledge production, *Futures*, 42(3): 254–264.

Halffman, W. (2009) Measuring the stakes: the Dutch planning bureaus, in P. Weingart and J. Lentsch (eds) *Scientific advice to policy making: international comparison*, 41–65, Opladen: Barbara Budrich.

Inderberg, T. and P. Eikeland (2009) Limits to adaptation: analysing institutional constraints, in W.N. Adger, I. Lorenzoni and K.L. O'Brien (eds) *Adapting to climate change: thresholds, values, governance*, 433–447, Cambridge: Cambridge University Press.

Jasanoff, S. (2004) Ordering knowledge, ordering society, in S. Jasanoff (ed.) *States of knowledge: the co-production of science and social order*, 13–45, New York: Routledge.

Jepperson, R.L. (1991) Institutions, institutional effects, and institutionalism, in W.W. Powell and P.J. DiMaggio (eds) *The new institutionalism in organizational analysis*, 143–163, Chicago, IL: University of Chicago Press.

Kuks, S.M. (2009) Institutional evolution of the Dutch water board model, in S. Reinhard and H. Folmer (eds) *Water policy in the Netherlands: integrated management in a densely populated delta*, 155–170, Washington, DC: RFF Press.

Latour, B. (2005) *Reassembling the social: an introduction to actor-network-theory*, Oxford: Oxford University Press.

March, J.G. and J.P. Olson (2006) The logic of appropriateness, in M. Moran, M. Rein and R. Goodin (eds) *The Oxford handbook of public policy*, 689–708, Oxford: Oxford University Press.

Noorderzijlvest (2012) *Droge Voeten 2050 waterschap Noorderzijlvest: plan van aanpak*, Groningen: Waterschap Noorderzijlvest.

Phillips, N., T.B. Lawrence and C. Hardy (2004) Discourse and institutions, *The Academy of Management Review*, 29(4): 635–652.

Pielke, R. (2007) *The honest broker: making sense of science in policy and politics*, Cambridge: Cambridge University Press.

Rayner, S. (2003) Democracy in the age of assessment: reflections on the roles of expertise and democracy in public-sector decision making, *Science and Public Policy*, 30(3): 163–170.

Rayner, S., D. Lach and H. Ingram (2005) Weather forecasts are for wimps: why water resource managers do not use climate forecasts, *Climatic Change*, 69(2–3): 197–227.

Scott, W.R. (2008) *Institutions and organizations: ideas and interests*, 3rd edn, Los Angeles, CA: Sage.

Turnhout, E., M. Stuiver, J. Klostermann, B. Harms and C. Leeuwis (2013) New roles of science in society: different repertoires of knowledge brokering, *Science and Public Policy*, 40(3): 354–365.

Vink, M.J., D. Boezeman, A. Dewulf and C.J.A.M. Termeer (2013) Changing climate, changing frames: Dutch water policy frame developments in the context of a rise and fall of attention to climate change, *Environmental Science & Policy*, 30: 90–101.

Yanow, D. and P. Schwartz-Shea (2006) *Interpretation and method: empirical research methods and the interpretive turn*, New York: ME Sharpe.

6 Governance of climate adaptation in Australia

Design charrettes as a creative tool for participatory action research

Rob Roggema, John Martin and Lisa Vos

Introduction

Planning for climate change is a complex task, often carried out by well-educated people working in sophisticated institutions. It is a task that requires knowledge about cohesive topics, different spatial scales, and overlapping interests. Research evidence suggests that successful adaptation is more likely if those impacted by extreme climatic events are involved in the planning for adaptation (Beer *et al.*, 2012; Nisbet, 2009). This involvement can be shaped in many ways. A design approach, as discussed in this chapter, combines a creative way of tackling the need for climate adaptation at community level, e.g. the actors that are impacted, with the opportunity of interaction and exchange between the involved participants, making social learning a key feature of these processes. We acknowledge social learning as especially applicable to wicked problems such as climate adaptation. Wicked problems such as climate change are characterized by uncertainty and ambiguity; they do not have a single, definite answer (Rittel and Webber, 1973). Social learning is important not only to reduce uncertainties, but also to develop consensus about problems and solutions among multiple actors.

In this chapter, we present the design charrette, an approach to planning which gives local communities the opportunity to become more climate resilient through collaborative design. A design charrette is defined as:

> a collaborative design and planning workshop that occurs over four to seven consecutive days, is held on-site and includes all affected stakeholders at critical decision-making points.
>
> (Lennertz and Lutzenhiser, 2006: 3)

In this chapter, we illustrate the influence that social learning through charrettes may have in communities and the way this contributes to more adaptable spatial futures.

In general, design is seen as a process of creating options, rejecting parts, reconstructing, and shaping, and is therefore fit to deal with wicked problems (De Jonge, 2009). In collaborative design processes, participants are invited to contribute to, and examine, the problem from various perspectives. In particular, the traditional roles of academics, citizens, or policymakers often prevent them from viewing a process from other perspectives. The participation of these people in the collaborative process of the design charrette made it possible to change roles and act in a different role than usual. In order to collaborate on climate adaptation planning and design, individual participants need to be able to understand one another. If in ordinary meetings with likeminded people there can already be a problem, imagine what level of communication is required when people are participating in processes with participants from many different backgrounds. Therefore, in our research project the participants were forced to leave their traditional roles of science and practice and take on a role to which they are usually unaccustomed.

Design charrettes (Condon, 2008; Lennertz and Lutzenhiser, 2006; Lindsey *et al.*, 2009) are a practical way for swapping traditional roles. To assist in dealing with wicked problems and rapid change, design charrettes comply with a set of conditions (Mandell and Steelman, 2003):

- *perception of the problem*: participants need to feel the urgency to deal with the problem;
- *intensity of linkages, existing or developing during the operation*: people need to be able to connect by means of their existing social networks or develop new connections;
- *breadth of effort*: is participation limited or comprehensive? participants need to be 'up for it', meaning they are really committed to participate and share the best of their expertise;
- *complexity of the purpose*: the aim is to tackle complex issues, but these need to be presented as relatively simple tasks, executed in a step-by-step manner; and
- *scope of effort*: status quo or system change? the objective is to identify fundamental change and this needs to be clear before the charrette starts.

Research question

How does changing roles in design charrettes contribute to social learning in regional planning processes?

Structure

This chapter starts with a description of the design-led methodology. We then discuss the advantages of taking on new roles as both scientific and practising communities. Following this, we illuminate the background of social learning and climate adaptation. The reason for the use of design-led charrettes as the planning process is clarified. Then the research findings are presented by addressing questions such as: How is the process of a design charrette prepared, executed, and evaluated? What are the typical results of a design charrette? What role do the researchers play during the design charrette? The chapter ends with a discussion of the findings and conclusions.

Methodology

The action research methodology used in this research project has been designed according to the design charrette approach and includes the principles of social learning. It consists of nine steps:

1   Prepare complex, mostly climate-related, information in a concise, understandable, and visible way.
2   Prepare and programme the charrette and select the right participants. The charrette programme in our cases lasted for two days and consisted of an introductory welcome with presentations of concise knowledge and prepared research outcomes (two hours), identification of aims and objectives (one hour), for instance in a 30–30 exercise,[1] sketch sessions at different scales (seven hours), 3D-modelling with plasticine (three hours), an appraisal of the outcomes (one hour), and final presentation of the results (two hours). The participants were recruited together with the local authorities or local strategic partners and community organizations. These groups understood the community very well and could advise about who the best people were to invite. These local stakeholders consisted of citizens seen as key players in the community, representatives of community groups (for instance green/ecology groups, neighbourhood representatives, local business groups, renewable energy proponents), local government employees and politicians, local business owners, and experts in the local environment (water, ecology, landscape). These local people were combined with academics in the fields of planning and design, climate science, experts in participatory planning processes, state representatives, and students of climate studies and landscape architecture.
3   Define the type of action research or the nature of the role of the researcher during the charrette. In this research project, the action research level chosen is participatory observation in combination with experimentation, the action research type is participatory action research, and the action research depth is co-production (see Huntjens *et al.*, Chapter 2, this volume). In this case, the researchers took part as fully integrated participants in the charrette and were an integral part of the practical process, whereas the charrette leader took on a distant role. The leader collected data by observing the participants, conducted many seemingly unstructured talks, and interviewed a selection of participants afterwards.
4   Define new roles for the participants in order to reshape the perspectives of academics and practitioners alike. In order to identify and distinguish new roles, a science–practice model is used in which the academic and public domain mutually influence each other and interact (Figure 6.1).
5   This leads to specific, alternative roles for academics and practitioners in each of the process steps during the design charrette. Before the start of the charrette, the researcher's role is that of an artist, translating his/her research into visual and iconic images. At the beginning of the

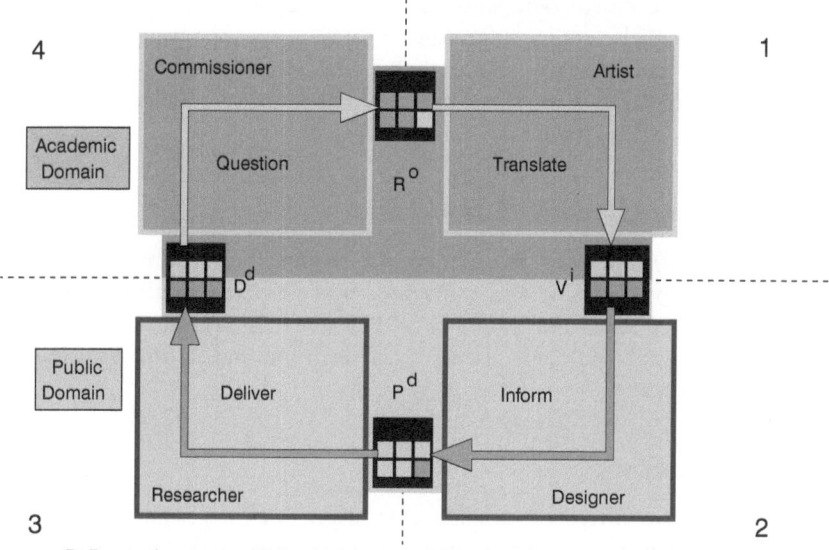

R° Research outcome; Vⁱ Visual information; Pᵈ People's design; Dᵈ Design data

Figure 6.1 Alternative model of academic–public interference (based on Roggema, *et al.*, 2013)

charrette, the researcher acts as a communicator and presents his/her research output as an artist. At the same time, the practitioners and local stakeholders become researchers, investigating the research data and the history, present, and future of the study area. During the sketching phases, both the researcher and the practitioner become designers, whereas the traditional design professionals play a modest and distant role in the sketching exercises. In the latter phases, all participants take on the role of builders and artists, shaping a future plan in 3D using plasticine. After the charrette, the results need to be researched. Here, there is a special role for the design professionals to carry out research by design. The charrette results are sophisticated and further designed and tested. Out of this new research, questions are identified. Finally, these research questions are adopted by researchers and transformed into new research proposals, which are adapted to the demands of practice and suitable for funding. In this final phase of the process, the researcher places him/herself in the role of a commissioner.

6 Execute the design charrette process. For each case study, two consecutive charrettes were executed of two days each.

7 Identify the success factors of the design charrettes. In this case, this was done by analysing the evaluation material, the personal feedback collected from participants, and the leaders' own experiences of the charrette.

8 Reflect on the role and position of action researchers and reflect on the new roles of scientists and practitioners by conducting a limited number of well-focused interviews with participants after the charrette.

9   Appraise the design outcomes, and reflect on and evaluate the benefits
     and results of the design charrettes by participants appraising the results at
     the end of the charrette and analysing the evaluation forms and personal
     feedback received from participants by the charrette leaders in informal
     conversational interviews.

Social learning and climate adaptation

New (potential) roles for those involved require new learning, especially
for climate adaptation, and this learning is best practiced in a social learning
environment (Fünfgeld and McEvoy, 2011). We first introduce the climate
adaptation context and then discuss social learning and its use in a climate
adaptation context.

Climate adaptation

With reference to Chapter 2, climate adaptation – defined as an adjustment in
natural or human systems in response to actual or expected climatic stimuli or
their effects – moderates harm or exploits beneficial opportunities (IPCC, 2007)
and is deemed necessary as the effects of greenhouse gas emissions stay with us
for a long period (IPCC, 2001). Nowadays, smooth, moderate climate change
is being replaced by more extreme weather events, increasingly occurring by
surprise (Steffen *et al.*, 2013). Moreover, the change is non-linear (Jones, 2010,
2011). This requires a more flexible planning methodology, as proposed in the
swarm planning approach (Roggema *et al.*, 2012a, 2012b), which emphasizes a
shift from the current technical to an organic paradigm, in which dynamic
open-ended solutions are possible (Broess, 2012; Brouwer, 2011).

Climate adaptation is framed in between a bottom-up and a top-down
approach, referring to social (with a focus on the past and on the present) and
physical (with a predominant focus on the future) vulnerability, respectively
(Dessai and Hulme, 2004). Despite the fact that communities are seen as a
crucial factor in these definitions, the communities' role in practice is weak.
Record-breaking weather across Australia (Steffen *et al.*, 2013) has intensified
the need for innovative adaptation strategies, which must be implemented in
local and regional land-use planning (CfES, 2012). Apart from adjusting our
built structures, we also need to adapt ourselves (Pijnappels and Dietl, 2013).
Community action is key to promoting adaptation (CfES, 2012). We need to
start looking at citizens as scientists. Country towns in particular face pressures
from climate extremes in addition to economic and population decline. Here, the
community *itself* is the greatest asset and source of potential success in climate
adaptation (Beer *et al.*, 2012). In many cases, adaptation is not tangible, e.g. not
visible or recognizable to the community, or it is hidden in measures taken in
other sectors (Pijnappels and Dietl, 2013). If the results of adaptation policies are
not experienced or recognized, people perceive less need to become involved or

act. Therefore, approaches and methods that provide tangible results and involve the community – e.g. design charrettes – are required.

Adaptation should be seen as a continuous process of learning and institutional change responding to new knowledge and changing circumstances; this contrasts with the viewpoint that adaptation is the result of technical measures but depends on people (Fünfgeld, 2012). Solutions are then collaboratively determined using a variety of information from different perspectives following the characteristics of adaptive governance (Swanson and Bhadwal, 2009), and adaptation becomes a social learning process (Fünfgeld and McEvoy, 2011; Horstmann, 2008). In our research project, these social learning principles have been applied to improve planning for climate adaptation.

Social learning

When (eco-)systems are managed for stability, they tend to end up on a pathway of turbulent change (Peterson *et al.*, 2003). The more emphasis is placed on managing towards a stable, unchanged situation, the more an ecosystem will develop a drive for change (Peterson *et al.*, 2003). This means that the opposite of the objective is realized. What has been concluded about the functioning of an ecosystem can be easily translated to organizations or governance systems: when strict regulations are put in place aiming to keep the system stable, individuals in organizations will aim to evade these rules, eventually leading to change in the entire organization. In a research context, it is no longer satisfying to fulfil the traditional role as a researcher, and this leads to scientists becoming one of several actors in the learning and knowledge generating process, together with local groups, business leaders, governments, and others (Kates *et al.*, 2001). On the positive side, a crisis seems to trigger learning and opens up space for new management trajectories (Westley, 1995) that are better at surviving turbulent times through re-organization processes (Hamel and Välikangas, 2003). Social learning contributes to the ability of local communities to self-organize and hence to deal with sudden and rapid change.

The core conditions for effective self-organization (Figure 6.2) are defined as follows (Vos, 2013):

1   It requires a strong sense of common identity, meaning, or purpose. In addition, individuals need to be able to connect their own sense of meaning to this overall identity, either consciously or subconsciously. This is possible when actors are capable of changing their role, either literally or mentally. Participants in the planning process need either to understand other roles, or – better – to be able to act in another role.
2   Effective self-organizing systems have rules that govern the way individual agents relate and interact. These rules becomes codified or reinforced when they improve the system's chances of achieving its purpose or maintaining its identity. Coding and reinforcing the governing rules is much easier when

Figure 6.2 Self-organization

participants are not limited to acting within their well-known competences, but experience the process from a stretched perspective, out of their natural comfort zone. This can be enhanced when people are given a new role or position.

3   It requires agents (participants), physical objects, and information to flow freely from where they are available to where they are needed. Without free movement of information, it is impossible for self-organization to emerge, because the elements of a self-organizing system can only self-organize when its elements are free to move. This is true for agents (the participants), physical objects, and information. What is more, when the carriers of information – the participants in the process – are forced to switch roles, the flow of information will also be strengthened.

Furthermore, the greatest depth (Homan, 2001) of learning impact is needed for a local community to deal with uncertainty and the highest climate impacts: simultaneous collective and individual learning applied to adaptive challenges, while moving multiple times through the *experiential learning cycle* (Kolb, 1984), which explains how learning takes place. There needs to be concrete experience ('acting'), conscious reflection on that experience, and abstract conceptualization – which can consist of expert knowledge being taught, books being read, or other methods – after which a learner decides on new types of behaviour to experiment with in practice, and this again leads to concrete experience (Kolb, 1984), and so on. So, whereas traditional education focuses almost solely on transfer of expert

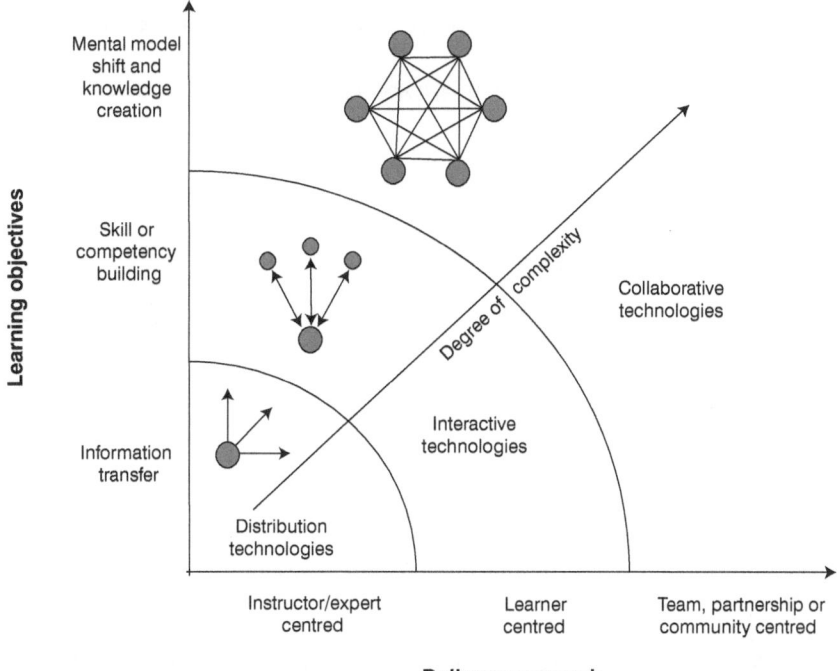

Figure 6.3 Learning objectives and delivery approach, adapted with permission from Keith McCandless, www.lberatingstructures.com

knowledge to learners, Kolb's cycle implies that no meaningful learning will occur from that activity alone. All four steps of the cycle need to encompass learning for actual learning to occur, in the sense that people are observed 'doing' differently in the context of their learning domain. Knowledge needs to make sense in the context of a learner's experience, of which they could only gain awareness by reflecting on that experience; and knowledge only makes sense if it then leads to new behaviour experiments in the context of the learner's practice. This is referred to as third-level learning (*breaking the frames*), which is distinguished from first-level learning (*framing*) and second-level learning (*reframing*) (Homan, 2001).

Learning levels consist of a combination of learning objectives ranging from information transfer, skills, and competency building to mental model shift and knowledge creation, and delivery approaches, ranging from instructor/expert centred via learner centred to team, partnership, or community centred (Figure 6.3). The highest degree of complexity is collaborative learning, which is needed to create novel solutions to wicked problems (Rittel and Webber, 1973). It involves collective learning and sense-making processes in which new shared views and mental models are created, which then serve as new and more effective *theories-of-action* (Argyris and Schön, 1974). Relevant knowledge is equally present in facilitators and learners and created in collaborative, self-organizing processes. To learn is to *break the frame*: share a diversity of multiple perspectives in order to generate a new, shared set of meanings in order to produce novel effective behaviour.

These highly complex learning processes imply an adjusted relation between citizens and policymakers and scientists. This is a process of co-creation through every phase of development and realization (Sanders, 2006): when a project is initiated (creating), designed (making), realized (adapting), and used (doing).

The conducted design charrettes

In this section, the findings from two case studies used to examine the design charrette process are described.

Background

Uncertainty about expected change complicates decision-making processes. This is especially the case in the governance of climate change in Australia. The combination of diverse climate change impacts, from floods to fire, drought, and severe storms, challenges policymakers as to how to respond. The current political apathy towards climate change means that this issue is not being addressed in a proactive manner because the Liberal-National Party, which won power in the 2013 federal elections, promised to repeal or dismantle all institutions and policy measures involved in climate change and clean energy (Parkinson, 2013). In this context, it is difficult to involve citizens and stakeholders in planning processes for climate change adaptation.

The Victorian State Government funded research to develop knowledge and identify strategies that would engage local and regional communities in decision making for climate adaptation. They were concerned that the various groups, in science and in policymaking, were focused only on their respective roles. The scientist seeks the truth about climate change and conducts his/her research, and the policymaker tries to embed research outcomes in the political arena. The citizen tries to deal with climate impacts in his/her daily life and needs a coordinated response from both science and policymakers.

The premise that knowledge is the sole prerogative of academia, that political choice is monopolized by policymakers, and that citizens are uninformed in both political and scientific dimensions of climate adaptation is questioned. The rationale for the state government supporting the research project, entitled 'Design-Led Decision Support for Regional Climate Adaptation', was to analyse whether a change in roles is beneficial for adaptive capacity in communities and whether those involved in planning processes would encounter novel ideas and outcomes as a result of this engagement, regardless of their specific knowledge about climate adaptation.

The tool used in this research project is the design charrette, which functioned as a real-life lab for exploring novel ideas on the impact of planning for, and anticipating, climate change. Design charrettes are an example of participatory action research (Baum *et al.*, 2006; see also Huntjens *et al.*, Chapter 2, this volume) in the complex arena of local/regional governance and climate change adaptation.

During the charrettes, researchers explored the 'real world' with communities, acquired new knowledge, and developed innovative visions. The participants involved in the design process are designers, climate scientists, local governance professionals, and local communities ('ordinary' citizens, 'local heroes', people representing action, environmental, or historic study groups). The level of involvement of all participants – decision makers and politicians, policymakers, academics, designers, citizens, and active community groups – was extremely high as everyone was an integral part of the entire process. Each participant was directly involved in every step and exercise of the design charrette and challenged to contribute to the action. Thus, the participants could not escape from contributing to the design process.

Case studies

In two cases, design charrettes were used to create local/regional design, enhancing a more climate adaptive future. In Bendigo – one of the largest country towns in Victoria – the main question was where new residential buildings could be best located, given the landscape's vulnerability to, mainly, bushfires and, to a lesser extent, heat waves and floods. The landscape surrounding Bendigo is very rich in forests, and this is the most important attractive factor for people wanting to reside here. At the same time, the forests increase the risk of bushfires. The charrette in Sea Lake, a small town of approximately 500 inhabitants, dealt with different questions. Here, the main climate impacts are drought and, occasionally, dust storms, heat waves, and minor flooding. Given the focus of the town on producing grain, the design process needed to provide new farming methods that could lead to continued agricultural production even under extreme drought conditions and also suggestions for a larger diversification of the economic base, to be profitable given future climate impacts.

Spatial designs

In both design processes, several spatial designs were produced, each giving spatial solutions for an improved climate adaptive future.

The Bendigo charrette delivered four integrative proposals, each of which addressed all the problems but each with a specific focus on a different climate problem, such as heat, fire, mitigation through renewable energy harvest, and water security, which was called 'The Scarcer the Water …',

In this design, the following climate adaptation measures were proposed (Roggema *et al.*, 2011b):

1 engage and encourage residential responsibility for capturing and using water efficiently;
2 introduce visible icons of adaptation, such as the proposal to use the Bendigo clock tower as a water meter;
3 protect and enhance natural waterways;

4 encourage self-sufficient agriculture through the production of food locally, for instance by creating market gardens close to the city;

5 anticipate drought by capturing stormwater before it reaches the creeks, save water in times of heavy rainfall, capture excess floodwater, apply the principles of water-sensitive urban design (Kay *et al.*, 2004; Wong and Brown, 2009), manage water seepages and evaporation of agricultural channels, and prevent discharges from treatment plants into streams;

6 adapt to droughts, heat, and heavy rainfall by reducing 'food miles' and introducing urban agriculture e.g. balcony gardens, rooftops, and so forth; and

7 anticipate heat-stress, droughts, energy blackouts, floods, and food scarcity by building self-contained housing types with integrated water collection, solar energy systems, and vegetable gardens.

From the Sea Lake charrette, three comprehensive proposals were gleaned, each with a specific focus: the growth of the town centre, the improvement of farming methods, and the diversification of economic functions.

In this proposition, the following measures are proposed (Roggema *et al.*, 2012b):

1 develop the arts in a specific precinct where ateliers can be housed in vacant buildings, with restaurants and a museum, screen movies using the old grain silos, and develop a planetarium in one of the old buildings;

2 develop the recreational potential, such as creating a Mallee Rally Museum and a Footy restaurant, creating new bike tracks around the lake, and organizing cyclo-cross races; improve the accommodation options in town and near the lakes and provide them with solar power and energy storage possibilities; develop an eco-lodge near Lake Tyrrell, where the fantastic night skies, stargazing, and the tranquillity of the lake can be experienced from a viewing platform at the edge of the lake;

3 improve education focusing on knowledge development in relation to salt water and energy harvesting, extend the existing secondary college, attract exchange students to the town, and use vacant buildings to accommodate students;

4 create an ecological community garden that harvests rainwater and energy; native vegetation should line all industrial estates;

5 collect salt and water, and store solar energy in the salt of Lake Tyrrell; and

6 diversify farming with new intensive farming of pigs, kangaroos, emu, and chickens.

Success factors

During these charrette processes, community members and local experts were mixed with academics, designers, and policymakers. As a group, they designed and modelled their desired future, which is expected to be capable of dealing with unforeseeable climate impacts. Whether this ambition was achieved was analysed in a process that took place partly during the design process and partly afterwards. At the end of the charrettes, the designs were 'judged' by the participants in terms of whether the proposals were seen as good or bad adaptation. Every participant received a set of yellow and green post-its and could place these (with explanation) on the designs. This provided us the information about the level of adaptation of the plans. Moreover, after the charrettes, the draft results were collected and refined in an 'internal' design process by university design researchers and students. This process not only improved the detail of the design, but also made a deep appraisal of the proposals possible. In this appraisal (Clune *et al.*, 2013), designs were analysed economically, ecologically, and socially, and a conclusion was drawn about their *adaptivity*.

The academics' role changed from that of provider of detailed and validated information to that of storyteller using the information, but explaining it in usable parts. This allowed the scientists to participate in the process as designers and think about gleaning research outcomes after the charrettes were concluded. The citizen-participants could change their role from being reactive to plans proposed by government or knowledge presented by scientists to that of expert, providing the group with new information that was seen as locally relevant and that would never have been put on the table in their traditional role. The decision-makers and the policymakers changed their role from that of explaining the policies and decisions to that of being a co-developer of plans, literally helping, supporting, and thinking along with the local community with ideas to improve adaptation. The designers also changed role. Instead of drawing the ultimate design on the map, they became observers of the process. Their role was to listen and understand the purpose and content of the drawings and models as proposed by the group. This way, they could not hijack the design process with their drawing skills, but could focus on translating the rough and unclear products into visually attractive images after the conclusion of the charrette. These new ways of collaboration and exchange between community members and the other participants, all with different interests, led to changed social constellations, both within the community and in the relationships of the external 'experts'.

The following were the success factors of the design charrettes as executed in Sea Lake and Bendigo (Clune *et al.*, 2012; Roggema *et al.*, 2011a; 2011b; 2011c; 2012b; 2013):

1   Role change proved to be extremely effective, as it allowed mutual insights to flourish, exploration of unknown new pathways, without preconditions or anyone taking a fixed position. The designers took on the role of facilitators, the researchers/academics became designers, the residents/local stakeholders changed their role to experts/knowledge owners, and the governmental representatives changed their traditional role to step into the shoes of citizens. Being or acting as a different person or role taught the participants to discover new insights, contribute these to the design process, and discuss

the different points of view with others (including their own role in daily life). This made the integration of ideas and insights, as well as opinions about a desirable future, much easier.

2   The role change allowed participants of different backgrounds, interests, and ages to build new social constructs. The mix of people present at the charrette developed collaboratively a shared perception of the problem, which then formed the focus for the entire charrette. The key players in a community, the charismatic group thinkers, and the specific experts in the field were critical in creating this success. These types of design processes were integrative, inclusive of a variety of solutions, and both intuitive and judging. This made the role change easier and accessible for all. In this sense, everyone was a designer.

3   Changing roles between adults and primary school children (which occurred only in the Sea Lake charrette) proved to be extremely fertile (Roggema et al., 2012a). Not only do these children have extraordinary ideas, they are also the people that actually will live in their own designs. Moreover, the exchange of roles between adults (being a child again) and children (suddenly the mighty deciders about the future) opened the constrained minds of adults in particular.

4   In these charrettes, the (re-)learning process appeared to be cyclical, including all elements of Kolb's (1984) experimental learning cycle, most probably as a result of the initiated role change. In each step of the design process, a specific phase of the cycle occurred. *Abstract conceptualization* was practised during the sketch phases, *reflection and observation* occurred during intermediate discussions and final presentations, and *concrete experiences and active experimentation* happened while the 3D-plasticine models were being built.

5   The free flow of information was an important factor before, during, and after the charrette. Moreover, when it was visual and 'just enough', it was easier to use and move. Too much information caused decision paralysis and too little produced bad proposals. Representing climate change visually (Barker, 2003) in the charrettes helped to conceptualize the problem, identify potential feedbacks, and improve communication between participants from different backgrounds and knowledge levels. The invitation and the challenge arose first and foremost when the charrette was started with a blank sheet. Everyone was invited to fill in their vision of the future on the previously blank paper, such as executed during the 30–30 exercise, sketching at different scales, and building a plasticine model. A strong ownership of both visuals and, more importantly, the visualized content was vested in the group of participants, because everyone contributed intensively to them. The execution of a tactile methodology, making use of drawing and building plasticine models, added to the depth of learning, because participants experienced different approaches to visioning, apart from talking and writing. The outcomes form a well-understood agreement. The drawings cannot be destroyed without the group's consent and function as a very strong commitment.

6   Some other findings, not directly related to the research questions, are: a good venue is essential, e.g. it is helpful if the venue is located at, or close to, the site, is convenient, has several rooms in which to break out, and is at a distance from participants' workplaces. The charrette leader only needs to create the space for participants to contribute (and keep the time), and to keep strictly to this task and nothing more, and it is essential to ensure the support of the local authorities.

Discussion

In retrospect, the many positives of the design charrette approach are accompanied by a couple of pitfalls. First of all, there is the risk of the researcher relapsing into his role as a researcher. It requires courage to open one's research and make it accessible to laymen. Especially in a design process, with unexpected moves, academic merits and reasoning might prevail over communicating results in a simple way. The second issue is that, for professional designers and planners, the results involving 'researchers in action' seem too quick and easy. This we find quite ironic. In regular planning processes, the results take much more time than in the two 2-day design charrette workshops. Decision makers find it hard to take them seriously and value the outcomes of these processes as much as those of their traditional research or planning processes. Third, the results are captured by elaborating the design results after the design charrette, but this follow-up does not guarantee the implementation of the plan. The question remains as to how to implement the outcomes. Finally, action research predominantly emphasizes the role of the action researcher in processes that are the object of study. It can be questioned whether action research should focus more on the contributions of the action researcher to the real-life process in order to make improvements. In the case of the design charrettes, the researchers contributed to these improvements. We are interested in the way these processes facilitate climate adaptation outcomes, how they are developed, understood, and owned by communities.

Conclusions

The change that participants' roles undergo in a charrette is beneficial for dealing with complex questions and when people with different backgrounds and knowledge levels are involved. The idea of asking researchers to become artists, designers to act as researchers, and citizens to design, enhances learning and mutual understanding and eases communication. The indirect advantage is that everyone involved takes on a contributing attitude, and this leads to designs that are supported by all.

The case studies have illustrated that using action research makes it possible to make complex academic knowledge available to local communities, using researchers as the medium. In this respect, it can be argued that action research is the key to transferring academic knowledge in an understandable and useable way to local practitioners and citizens. Moreover, this setting makes it possible to accept local expert knowledge as valuable empirical data in a research environment. In this respect, action research can be defined as a two-way, mutually enriching approach, deepening understanding of both scientists and non-academics.

The most important barrier to using action research as the methodology in climate adaptation research is the lack of acceptance of this type of research process by local government councils. The councils in our cases were concerned about the time investments, which they thought to be very time-consuming. They had trouble committing the relevant stakeholders and key players within the community and the council, and they were afraid that the outcomes might interfere with existing policies. In the case studies, we have seen that the main solution to this problem is always to connect the climate adaptation problem to a pressing and actual issue in the local situation. The problem of climate change is sometimes so overwhelming that it paralyses the daily practice of policymaking. Since the core business of many governance processes is to focus on daily issues that can be easily grasped, climate change issues may easily become a taboo topic amongst policymakers, and that is why connecting 'under cover' climate adaptation with these daily issues is so important.

The two cases illustrate that, when participants in a design process are encouraged to change their roles, social learning and learning experiences are amplified. In the two case studies, the appraisal of the many design proposals showed that these plans reflected a high level of adaptation in multiple senses – climatic, social, and ecological. However, it is not enough only to change roles. The design process needs to be well shaped, and the design charrette approach offers the environment within which role change and social learning can easily take place. The experiences in the two case studies illustrate that design charrettes can overcome everyday barriers, such as the 'closing-the-ranks' behaviour that occurs in many organizations but stands in the way of achieving innovative and ground-breaking solutions. When people participate in these design charrettes, they are confronted with inevitable requirements, such as group work, tactile techniques they have never used before, or only a long time ago, and quickly changing tasks that need to be completed within a pressing time schedule. This brings them, logically, out of their comfort zone. At this point, which coincides with them changing their role, consciously or unconsciously, they are suddenly exposed to new habits, knowledge, pathways, and people. This causes an openness and bonding within the group that makes unexpected outcomes possible. It is simply 'not done' to retreat to a safe comfortable position and, additionally, this is much less fun.

In the design charrette process, the results of climate change research are used as knowledge input in different ways and phases of the charrette: (1) as part of the design brief beforehand: abstract schemes and maps, representing complex climate data and predictions, but also the specific policy context; (2) during the charrette, climate knowledge plays a role in the process through participating climate scientists; and (3) afterwards in appraising and detailing the charrette results. Analysis of the evaluations showed that both scientists and practitioners were extremely satisfied with the results, with an average score of 8.7 out of 10. Climate knowledge found its way into concrete practice and could enhance innovative results, which most probably would not be possible in regular planning processes, and certainly not within a 2-day timeframe.

The design charrette produces tangible results in the form of more adaptive spatial plans and visualizes futures that are collaboratively developed using the expertise and ideas of citizens in combination with academic understanding of the problem. This approach is very useful in a local setting, because climate problems require local adaptation; and the ideas and – more importantly – the contributions of residents improve the value of the outcomes.

When academic researchers collaborate in the charrette process, they are forced to formulate their knowledge in a way that every participant can understand it. This makes it a very effective approach when a complex problem, the task of designing, and a mix of people, including experts from the community, form the ingredients of the planning process.

Note

1 The 30–30 exercise asks participants to look back 30 years in history and describe how technology, social aspects, and climate appeared, describe the present situation, and imagine how these themes would turn out 30 years into the future.

References

Argyris, C. and D. Schön (1974) *Theory in practice*, San Francisco, CA: Jossey-Bass.

Barker, T. (2003) Representing global climate change, adaptation and mitigation, *Global Environmental Change*, 13(1): 1–6.

Baum, F., C. MacDougall and D. Smith (2006) Participatory action research, *Journal of Epidemiology and Community Health*, 60(10): 854–857.

Beer, A., S. Tually, M. Kroehn, J. Martin, R. Gerritsen, M. Taylor, M. Graymore and J. Law (2012) *Australia's country towns 2050: what will a climate adapted settlement pattern look like?* Final Report for the National Climate Change Adaptation Research Facility, Nathan, Queensland: Griffith University.

Broess, H. (2012) *Onbedekt bloeiend*, Groningen: Academie van Bouwkunst, lezing Opening Krimpatelier, 7 February 2012.

Brouwer, J. (2011) *De eindeloze trap*, Doetinchem/Enschede: AfdH Uitgevers.

CfES (2012) *Climate change Victoria: the science, our people and our state of play*, Foundation paper one, Melbourne: Commissioner for Environmental Sustainability, The State of Victoria.

Clune, S., R. Roggema, R. Horne, S. Hunter, R. Jones, J. Martin and J. Werner (2012) *Design-led decision support for regional climate adaptation*, Final Report draft 1.0, Melbourne: VCCCAR and RMIT University.

Clune, S., R. Horne, R. Roggema, J. Martin and P. Arcari (2013) *Sustainability appraisals of design-led responses to climate adaptation*, policy brief, Melbourne: VCCCAR.

Condon, P.M. (2008) *Design charrettes for sustainable communities*, Washington, DC: Island Press.

De Jonge, J.M. (2009) Landscape architecture between politics and science: an integrative perspective on landscape planning and design in the network society, PhD thesis, Wageningen University.

Dessai, S. and M. Hulme (2004) Does climate adaptation policy need probabilities? *Climate Policy*, 4(2): 107–128.

Fünfgeld, H. (2012) *Local climate change adaptation planning: a guide for government policy and decision makers in Victoria*, Melbourne: VCCCAR and RMIT University.

Fünfgeld, H. and D. McEvoy (2011) *Framing climate change adaptation in policy and practice*, Melbourne: VCCCAR and RMIT University.

Hamel, G. and L. Välikangas (2003) The quest for resilience, *Harvard Business Review*, 81(9): 52–63.

Homan, T. (2001) *Teamleren: theorie en facilitatie*, Schoonhoven: Academic Service.

Horstmann, B. (2008) *Framing adaptation to climate change: a challenge for building institutions*, Discussion Paper 23/2008, Bonn: Deutsches Institut für Entwicklungspolitik.

IPCC (2001) *Climate change 2001: Synthesis report, A Contribution of Working Groups I, II, and III to the Third Assessment Report of the Intergovernmental Panel on Climate Change*, Cambridge/New York: Cambridge University Press.

IPCC (2007) *Climate change 2007: The physical science basis, working Group I Contribution to the Intergovernmental Panel on Climate Change Fourth Assessment Report*, New York: Cambridge University Press.

Jones, R. (2010) A risk management approach to climate change adaptation, in R.A.C. Nottage, D.S. Wratt, J.F. Bornman and K. Jones (eds) *Climate change adaptation in New Zealand: future scenarios and some sectoral perspectives*, 10–25, Wellington: New Zealand Climate Change Centre.

Jones, R. (2011) Planning with plasticine, retrieved from: http://2risk.wordpress.com/2011/11/30/planning-with-plasticine/ (accessed 30 November 2011).

Kates, R.W., W.C. Clark, R. Corell, J.M. Hall, C.C. Jaeger, I. Lowe, J.J. McCarthy, H.J. Schellnhuber, B. Bolin, N.M. Dickson, S. Faucheux, G.C. Gallopin, A. Grübler, B. Huntley, J. Jäger, N.S. Jodha, R.E. Kasperson, A. Mabogunje, P. Matson, H. Mooney, B. Moore, T. O'Riordan and U. Svedin (2001) Environment and development: Sustainability science, *Science*, 292(5517): 641–642.

Kay, E., T. Wong, P. Johnstone and G. Walsh (2004) Delivering water sensitive urban design through the planning system, in *Proceedings of WSUD 2004: cities as catchments, International Conference on Water Sensitive Urban Design*, 170–180, Barton, ACT: Engineers Australia.

Kolb, D.A. (1984) *Experiential learning*, Englewood Cliffs, NJ: Prentice Hall.

Lennertz, B. and A. Lutzenhiser (2006) *The charrette handbook. the essential guide for accelerated, collaborative community planning*, Chicago, IL: APA.

Lindsey, G., J.A. Todd, S.J. Hayter and P.G. Ellis (2009) *A handbook for planning and conducting charrettes for high-performance projects*, Washington, DC: National Renewable Energy Laboratory.

Mandell, M.P. and T.A. Steelman (2003) Understanding what can be accomplished through interorganizational innovations, *Public Management Review*, 5(2): 197–224.

Nisbet, M.C. (2009) Communicating climate change: why frames matter for public engagement, *Environment, Science and Policy for Sustainable Development*, 51(2): 12–25.

Parkinson, G. (2013) Tony Abbott sworn in, turns against climate programs, retrieved from: http://reneweconomy.com.au/2013/tony-abbott-sworn-in-turns-against-climate-programs-15946 (accessed 4 November 2013).

Peterson, G.D., S.R. Carpenter and W.A. Brock (2003) Uncertainty and the management of multistate ecosystems: an apparently rational route to collapse, *Ecology*, 84(6): 1403–1411.

Pijnappels, M. and P. Dietl (2013) *Circle-2 adaptation inspiration book. 22 implemented cases of local climate change adaptation to inspire European citizens*, Lisbon: Circle-2.

Rittel, H. and M. Webber (1973) Dilemmas in a general theory of planning, *Policy Sciences*, 4(2): 155–169.

Roggema, R., R. Jones, A. Soh, S. Clune, S. Hunter, A. Barilla, Z. Cai, J. Tian and J. Walsh (2011a) *City of Greater Bendigo design charrette I: the Report*. Melbourne: RMIT University, La Trobe University, Victoria University and VCCCAR.

Roggema, R., J. Martin and R. Horne (2011b) Sharing the climate adaptive dream: the benefits of the charrette approach, in P. Dalziel (ed.) *Proceedings 'ANZRSAI Conference'*, 281–292, Canberra, 6–9 December 2011, Lincoln: AERU Research Unit.

Roggema, R., J. Martin, R. Horne, R. Jones, S. Hunter, A. Soh, J. Werner and S. Clune (2011c) *Design brief design charrette I, City of Greater Bendigo*. Melbourne: RMIT University, La Trobe University, Victoria University and VCCCAR.

Roggema, R., R. Jones, S. Clune and D. Lindenbergh (2012a) *Sea Lake charrette, primary schools*, Melbourne: RMIT University, La Trobe University, Victoria University and VCCCAR.

Roggema, R., R. Jones, S. Clune and D. Lindenbergh (2012b) *Sea Lake charrette, dancing under the stars*, Melbourne: RMIT University, La Trobe University, Victoria University, and VCCCAR.

Roggema, R., J. Martin, P. Arcari, S. Clune and R. Horne (2013) *Design-led decision support: process and engagement*, Melbourne: VCCCAR.

Sanders, L. (2006) Scaffolds for building everyday creativity, in J. Frascara (ed.) *Design for effective communications: creating contexts for clarity and meaning*, 65–77, New York: Allworth Press.

Steffen, W., L. Hughes and D. Karoly (2013) *The critical decade: extreme weather*, Canberra: The Climate Commission.

Swanson, D. and S. Bhadwal (eds) (2009) *Creative adaptive policies: a guide for policy-making in an uncertain world*, New Delhi: SAGE.

Vos, L. (2013) Innovations in collaborative and community learning, in R. Roggema (ed.) *The design charrette: enhancing community resilience*, 35–60, Dordrecht: Springer.

Westley, F. (1995) Governing design: the management of social systems and ecosystems management, in L. Gunderson and C.S. Holling (eds) *Barriers and bridges to the renewal of ecosystems and institutions*, 391–427, New York: Columbia University Press.

Wong, T.H.F. and R.R. Brown (2009) The water sensitive city: principles for practice, *Water Science & Technology*, 60(3): 673–682.

7 Adaptive governance in practice

A learning approach based on
action research designed for the
implementation of climate adaptation
measures

*Gerald Jan Ellen, Corniel van Leeuwen,
Wiebren Kuindersma, Bas Breman and Frank
van Lamoen*

Introduction

As stated in Chapter 1, climate adaptation policy is relatively new. Therefore, it is not surprising that the implementation of climate adaption measures and policy has not received much attention in the governance literature (Dupuis and Knoepfel, 2013; Hovi *et al.*, 2009). In this chapter, we aim to focus on this subject. We build on the general implementation literature, which recognizes implementation as highly complex and difficult (Crosby, 1996; Sabatier and Mazmanian, 1980). Majone and Wildavsky even described implementation as 'the continuation of politics with other means' (1978: 175). In comparison with 'regular' implementation, climate change adaptation could be perceived as even more difficult as it is characterized by a long-term horizon with high levels of uncertainty (on the impacts of climate change). The scope of the problems and solutions can often be 50 years or more (e.g. Adger *et al.*, 2005; Hallegatte, 2009; Underdal, 2010, see also van Buuren *et al.*, Chapter 1, this volume).

Dealing with such long-term policy problems and fundamental uncertainties calls for an adaptive governance approach (Adger *et al.*, 2005; Cooney and Lang, 2007). Adaptive governance has emerged as a response to the increasing vulnerability of socio-ecological systems and is designed to deal with uncertainty and changed preferences in both the physical and the social system in an integrated and multidisciplinary manner (Folke *et al.*, 2005; Gunderson, 1999; Olsson *et al.*, 2006). It refers to a process of continuous monitoring, gaining knowledge about the functioning of actual strategies, and improving strategies in response to the changed context (Pahl-Wostl, 2007). The literature on adaptive governance describes the concept as promising in effectively dealing with unexpected developments (both social and physical), new insights, and changing circumstances (Termeer *et al.*, 2010).

However, the implementation of climate adaptation measures is confronted with several barriers, such as: lack of financial resources, unclear distribution of responsibilities and tasks due to an institutional void, discussions about the distribution of costs and future benefits, lack of trust between participants, and conflicting timescales and interests (e.g. Amundsen *et al.*, 2010; Bai, 2007; Biesbroek *et al.*, 2011; Giddens, 2009; Uittenbroek *et al.*, 2012). As a consequence of these barriers, the actual process of transforming towards a new – more adaptive – regime is confronted with challenges and problematic trade-offs (Van Buuren *et al.*, 2014; Lee, 1999). One of these challenges is that an adaptive governance approach puts specific demands upon how interventions are implemented, maintained, and monitored (Wallington and Moore, 2005).

Our current instruments and arrangements for implementing adaptation measures seem to be designed on the basis of the conventional, static, and predictable ways of policy implementation, and thus need to be transformed when the intention is to implement climate adaptation measures. The challenge is to accept dynamics and uncertainty, to be prepared for unexpected feedback patterns (Folke *et al.*, 2005; Haasnoot *et al.*, 2013), and to be able to respond to unexpected developments. This demands an approach based upon continuous learning, experimentation, broad (processes of) participation, and flexibility. In other words, in order to deal with the flexibility and uncertainty inherent in climate adaptation strategies, arrangements for implementation should be flexible and adaptive.

In this chapter, we describe a specific research project that focused on the implementation of adaptation measures (Ellen *et al.*, 2014). The aim of this project was not only to analyse the implementation of adaptation policy in practice, but also to facilitate the implementation of adaptation policy. Within this project, we were interested in flexible policy arrangements in particular. Policy arrangements can be described as combinations of different (legal, financial, and organizational) policy instruments and monitoring/reflection. As Gupta *et al.* (2011) state, the enabling of adaptive governance requires governance arrangements that facilitate this adaptability by being flexible and capable of adjustment. Consequently, the basic assumption of this project was that effective policy arrangements for climate adaptation needed to be flexible, thus implying that policy implementation should be characterized by:

- high levels of stakeholder participation (Edelenbos *et al.*, 2013; Warner *et al.*, 2013);
- agreement based on trust rather than on control (Bradach and Eccles, 1989; Edelenbos and Eshuis, 2011; Gambetta, 1988; Ring and Van de Ven, 1994); and
- possibilities to alter policy objectives and instruments along the way based on continuous (participatory) monitoring/reflection (Cundill and Fabricius, 2009; Plummer and Armitage, 2007; Plummer *et al.*, 2013).

The project was designed as a combination of case study research (Yin, 2003) and action research. The case studies focused on identifying (and understanding) the practice of policy implementation in the field of climate adaptation. The action research focused on designing a survey and a serious game as a tool for implementation. The survey and the game were developed based on: (1) general notions derived from the governance implementation literature, (2) the outcomes of the empirical case studies, and (3) interactive sessions with stakeholders in a specific case study (see next section).

The main research questions of this study (and this chapter) are: (1) What are the major (main) barriers to the development of flexible arrangements in order to be able to deal with the implementation of climate adaptation measures? (2) In what way can action research add to the insights on these barriers and can it help to overcome these barriers?

Following the typology of approaches to action research presented in Chapter 2 by Huntjes *et al.*, the action research conducted can best be described as action science, as the inquiry also aimed to identify 'the theories that actors use to guide their behavior' (Reason, 2003: 273). In this case, this refers to the theories used by the actors to formulate implementation arrangements for a regional water adaptation plan. The level of stakeholder involvement in this action research can best be described as reflection (by the researchers on the practitioners' actions).

This chapter continues with a description of the case study that was the object of our action research (the regional Deltaplan of the High Sandy Soil Areas). The next section elaborates on the methodology of our study and the combination between traditional case studies and action research (in a specific case study). In the fourth section, we describe and discuss the empirical results of the study. The chapter concludes by presenting answers to the main research questions and a reflection on our methodology.

The Deltaplan High Sandy Soil Areas case

Generally, climate adaptation in the Netherlands is associated with rising sea levels and flooding. Regionally, however, water scarcity and drought have also to be considered possible effects of climate change in the country. This will have consequences for the freshwater supply in certain parts of the Netherlands and could have damaging effects on Dutch agriculture and nature. Drought is expected to be a problem in particular for the sandy soil areas in the south and the east of the Netherlands. However, this does not mean that flooding is not (simultaneously) an issue in these areas. Extreme rainfall and flooding in certain periods are expected to go hand in hand with drought in other periods.

In 2009, a project was initiated in the south of the Netherlands to deal with the consequences of climate change in general and the effects of droughts in particular (see Figure 7.1). This project started as an initiative of 13 governmental and non-governmental stakeholders. Among the stakeholders were five regional water boards, two provinces, two farmers' unions, the state forestry service, Rijkswaterstaat, and two drinking water companies. This initiative was named the Deltaplan High Sandy Soil Areas (DHZ). DHZ (DHZ, 2014) can be seen as a regional specification of the National Delta Programme (2010–2014), although it is not officially part of it. See Box 7.1.

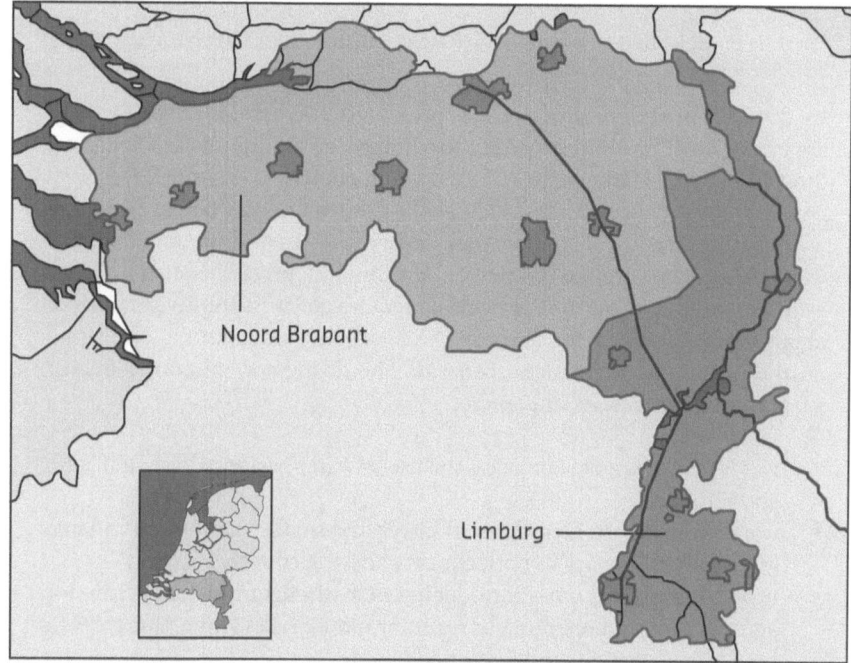

Figuur Ellen, Gerald Jan 14RS

Figure 7.1 Location of the Deltaplan High Sandy Soil Areas case

The DHZ has adopted similar methods of operation as the National Delta Programme. It started with an analysis of the (possible) consequences of climate change (in different scenarios). This was followed by designing possible strategies to deal with these consequences (phase 2-2012) and with a selection of promising strategies (phase 3-2013) and finally – currently taking place – the preferential strategies (phase 4-2014).

Our involvement in this project can be situated at the beginning of phase 3. In the first two phases, most time and effort had been spent on studying the physical system and technical design of a whole range of possible climate adaptation strategies and specific measures. However, this inventory of possible strategies and measures failed to address governance questions such as: Who will implement these measures? Which policy instruments can be selected for their implementation? Who will pay for it? How can we cooperate in an intelligent way? Who can or should contribute? Who is responsible? Which agreements can we make concerning implementation of the strategies?

In the next section, we elaborate on our research methodology.

Methodology

As already indicated, our research methodology consisted of a combination of traditional methods (case studies and a literature review) and action research

Box 7.1 Possible climate adaptation strategies within the Deltaplan High Sandy Soil Areas case (DHZ, 2014)

In 2012, the project group DHZ presented the results of phase 2: a document with possible strategies for climate adaptation with a focus on drought (Royal Haskoning, 2012). The document contains suggestions both for large-scale measures, such as the construction of one or two large water storage facilities, and for small-scale measures to promote numerous water basins on private properties (farms and private houses). Their estimate was that these measures would require government investments of almost one billion euro for the next 25 years. These calculations were based on the most extreme climate scenario. The document contains different adaptation strategies. Examples are:

- water conservation – limiting the use of water by agriculture, industry, and citizens;
- water storage – storing water in larger and smaller basins and in large and small waterways in both the city and the countryside; and
- new forms of land use – introducing new (drought resistant) crops for agriculture and more drought resistant nature types (from pine trees to broad-leaved trees), and so on.

In general, these measures have been formulated in rather technical terms and the way to implement them is indicated (either by rules, subsidies, or communication instruments); the actors involved and the division of costs and benefits have not been addressed. Some of the measures imply drastic and highly controversial policy decisions, such as the suggestion that water boards could intervene in farmers' individual freedom to choose the crops they want to grow. From a flexible arrangement perspective, the choice to elaborate only the most extreme climate change scenario and not several scenarios is remarkable (and perhaps questionable).

applied within the context of the DHZ case (see also previous section). In this section, we elaborate on our methodological choices and the way we have tried to connect the different methods.

Traditional method: case studies

In the first phase of our research, we focused on the use of traditional research methods. The reason for this was that we wanted to gain insights into past experiences in applying (flexible) policy arrangements in the implementation of adaptation policy and what hampered this implementation.

To do so, we chose the case study method (Yin, 2003). The main advantage of case studies is that they enable participants to tell their stories (Crabtree and Miller, 1999). Through their stories, the participants are able to describe their views of reality, and this enables the researcher to better understand the participants' actions (Lather, 1992; Robottom and Hart, 1993). Furthermore, with our project we wanted to generate insights into the 'how' and 'why' questions concerning the implementation of current adaptation practices; this, according to Yin (2003), can be answered through the use of case studies.

We selected nine cases by strategic sampling (Flyvbjerg, 2006). Eventually, we chose nine adaptation projects that had been, or were being, implemented at the time. Other selection criteria included: a certain spread over different adaptation issues (drought, flooding, and so forth), a geographical spread over the country, and (an expected) variety of implementation arrangements. The data collection was conducted by secondary analysis of existing research data (including 51 previously conducted interviews with stakeholders), additional interviews (22), and desk studies of documents (project and policy documents, formal and informal contracts between stakeholders, reports, newspaper articles, and websites).

Action research in the DHZ case

After the case studies, the research project started to focus on the DHZ case. We approached this case with multiple goals. On the one hand, we wanted to add insight from the case studies and literature to the ongoing policy process within this actual case; on the other hand, we wanted to apply and develop our ideas on the possibilities for, and barriers to, flexible arrangements that we developed from the case study analysis.

In order to combine our two objectives, we decided to develop two tools: a survey for individual actors and a serious game to apply in a group setting. The development of these tools and their application in the case study setting went hand-in-hand. We used prototypes of the survey and the serious game in different sessions with the case stakeholders. After those sessions, we used our experiences and the actors' reflections to adjust and improve these tools. This approach can best be described as a 'learning by doing' (Gibbs, 1987) approach to action science (Argyris, 1996; Reason, 2003), which aims to identify the theories that actors use to guide their behaviour. Key characteristics are reflection on action strategies (single-loop learning) and identification of mechanisms that underlie action (double-loop learning) (Huntjens *et al.*, Chapter 2, this volume). The survey was aimed mainly at single-loop learning and the serious game at double-loop learning.

The main objective of the survey was to serve as a kind of 'self-diagnosis', or self-regulated learning (Nicol and Macfarlane-Dick, 2006) or natural learning (Armstrong, 1979) for the DHZ project group. In an online survey, individual members of the project team were asked about the following subjects (based on the theoretical framework as previously described and as shown in Figure 7.2): the perceived objectives of the project, the level of political commitment, the desired distribution of costs and benefits, the legal, economic, and societal conditions

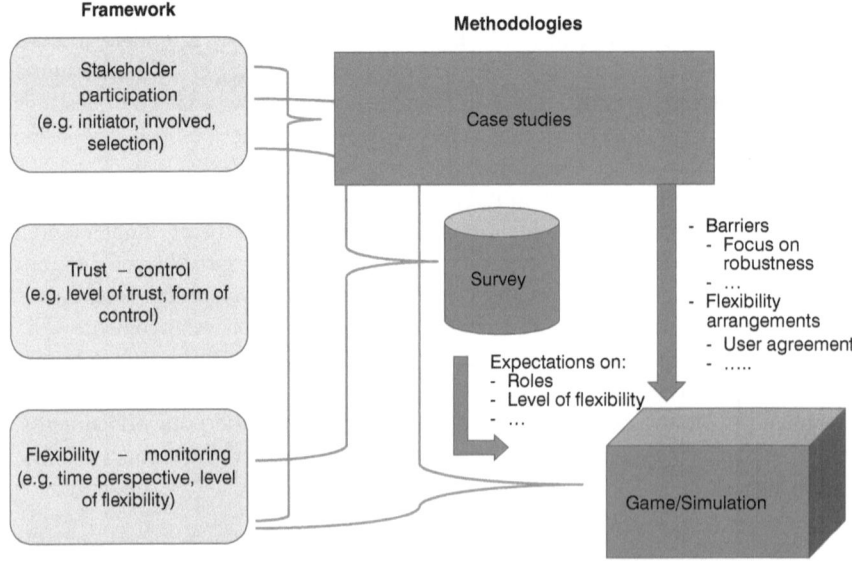

Figure 7.2 Overview of the research design

of implementation, and the desired flexibility of policy implementation. The idea behind the survey was that we could discuss the individual ideas of the project members and compare them, thus identifying differences and similarities in perceptions about DHZ and how to go about implementing the adaptation measure that would come out of it. By doing so, we wanted to create awareness on adaptive governance issues. The survey was used three times in different forms. The first time, it was used with the DHZ project team, the second time, as a reflection tool during a DHZ spatial design session, and finally in a smaller setting focusing on a pilot of a specific adaptation measure used in DHZ.

The second methodology that we used and developed during these interactions was a *serious game* (see Box 7.2). To define a serious game, we refer to Abt (1975: 9):

> serious games are games [which] have an explicit and carefully thought-out educational purpose and are not intended to be played primarily for amusement.

In our case, the term 'educational' can be interpreted broadly, e.g. sensing, thinking, and doing; these are three different phases of learning according to Kolb (1984). Since World War II, games have been used in the public sector to support policy- and decision-making processes (Duke & Geurts, 2004). More than 60 years of experience shows that simulation games have educational, training, policy, and research purposes. Games have proved useful for experimentation and learning (Mayer, 2009).

Box 7.2 FLEXAR: game description

Setting of the game

The game, which lasts about two hours, is best played with a maximum of 20 people. The smaller the group, the more space for discussion and interaction. The game was designed with a facilitator in mind.

Challenge

To meet the Reality requirement, the game uses a (fictional) climate adaptation challenge dealing with flooding due to heavy rainfall.

Different roles

To enhance the Play aspect of the game and to challenge the participants to think outside their usual patterns, the following roles were created: Market, Government, and Society. These are loosely based on the individualist, hierarchical, and egalitarian perspectives from cultural theory (Thompson et al. 1990). During the game, the participants work in pairs (or a multiples thereof) in the same role. The reason for this is that, by doing so, the participants can discuss their role, and this in turn helps them to 'get into' their role.

Gameplay

In the first of three rounds, the participants have to rate – from the perspective of their role – four criteria for the agreement that might be reached. These four criteria, a simplified version of Mees *et al.*'s (2014) criteria:

- Reaching goals: certainty that objectives are met;
- Flexibility: the ability to change agreements in different dimensions (planning, resources, and so forth);
- Enforcement: the extent to which agreements are enforceable;
- Participation in implementation: involvement of stakeholders in the implementation of measures and agreements (see Figure 7.3).

In the second round, the participants are asked to implement a local adaptation measure which, by means of small sluices, keeps the water in canals and agricultural areas – thus preventing the water from going downstream: keeping the lower-lying city dry. To do so, they are presented with factsheets of 20 policy instruments and three types of monitoring. They are then asked to select two policy instruments plus one form of monitoring. Furthermore, they are asked who should take the initiative for

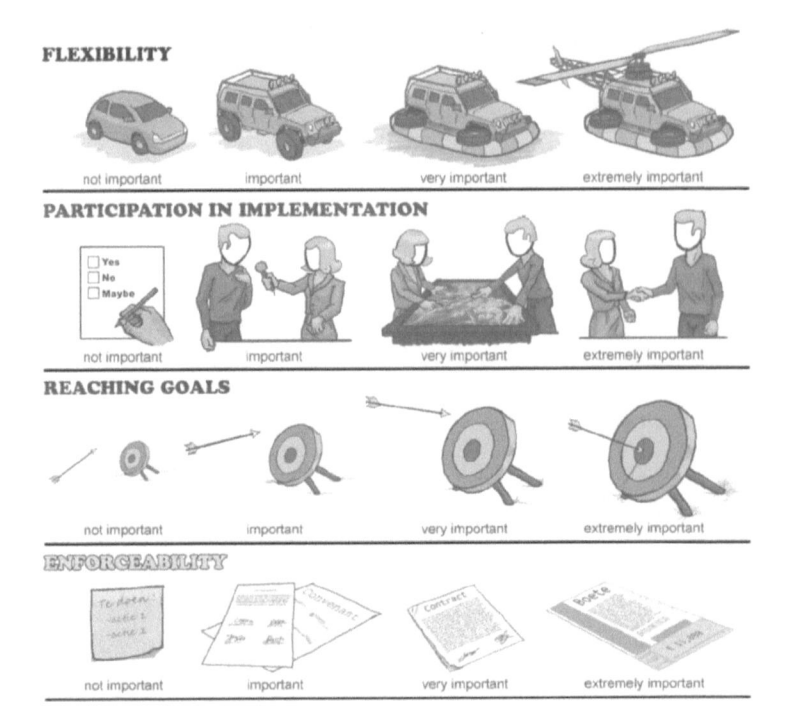

Figure 7.3 Four design criteria for the arrangement to be made in the game

the funding of this measure and which (group of) stakeholder(s) is (are) the main beneficiary. The differences in the choices between the groups are discussed and reflected upon; for this, we use the factsheets of policy instruments and monitoring as an important input.

The third and final round gives the participants new information telling them that rainfall will increase and that the small-scale measures will no longer deliver the required reduction of water runoff. Therefore large-scale water retaining areas have to be developed. Furthermore, they now have all the factsheets of policy instruments. On the basis of this new information, they are asked to reconsider their original choices concerning policy instruments, form of monitoring, funding, and main beneficiary.

The reasons for choosing a serious game were twofold. First, from practice, the DHZ case stakeholders suggested the development of such a game themselves, as they had positive experiences with serious games in the past. Second, games can contribute to: *stimulating creativity* and *commitment to action* (Bekebrede, 2010; Duke and Geurts, 2004). Creativity can be stimulated by immersing stakeholders in a game setting, thus causing players to leave their routines behind – as far as possible. This new environment encourages creativity. Commitment to action is relevant because serious games can be used to introduce or test new policies or interventions, to convince participants of the need for the intervention, to introduce the intervention approach, and to show the roles of the participants in the intervention (Bekebrede, 2010: 66). (For a more elaborate description of the use of serious gaming in science and practice, see Driscoll and Lehmann, Chapter 8, and Schenk and Susskind, Chapter 9, this volume.)

To design the 'Flexar' game, we used the triadic game design approach (Harteveld, 2010). This design theory is one of the better-documented design theories. The triadic game design method is based on the concept that a serious game should contain a balance between the trinity of *Play* (is it fun?), *Meaning* (can I retrieve any value from it?), and *Reality* (can I use it in the real world?).

Empirical research

At the start of this chapter, we argued that the implementation of adaptation measures requires flexible arrangements and that the development of such flexible arrangements is often hindered by barriers. The first goal of this project was to identify these barriers through the analysis of case studies. The second goal was to design (and apply) a tool for designing flexible policy arrangements in practice.

Case studies

From the case studies, we have been able to identify the following important barriers to flexible arrangements.

A first barrier to flexible arrangements is policymakers' preference for robust (one-time) interventions. In cases where climate scenarios and challenges concerning climate change are long-term in nature, arrangements and adaptation measures considered as solutions are of a short-term nature, that is, pilots or experiments that will last one to two years with very little room for flexibility. From the case studies, we assume that there is a strong preference for robust infrastructural measures above long-term recurring measures.

A second barrier is that – despite the long-term character of adaptation goals – detailed contracts are the preferred *modus operandi*. For example, subsidies and compensation for possible damages are frequently part of these agreements. Usually, these are meant to secure agreements and are not meant to enable a long-term dialogue on the adaptation of climate adaptation measures. Even though the level of detail was high, no clauses were incorporated that asked for a review of the agreement if external factors required that. From the case studies, it appears that there is some reluctance to using arrangements aimed at guiding adaptability.

This barrier can also be reframed however. In general, distrust between actors is an important driver for making detailed agreements and contracts between different stakeholders. Some of the case studies showed, however, that a high level of trust between stakeholders could have the same effect. Cases that show high levels of trust between stakeholders also show a relatively high tendency for stakeholders to commit to detailed agreements. Behind this, there could be an expectation among the stakeholders that these detailed agreements could be revised rather easily in the event of changed circumstances.

A third barrier is the existence of one very powerful actor. This implies that such an actor (often a governmental organization) can enforce adaptation solutions on other actors without too much opposition. This barrier also prevents other actors from coming on board as possible shareholders. Often, governmental organizations seem to be the first port of call when it comes to climate change and adaptation measures. This, however, reduces opportunities for private initiatives and public–private partnerships.

A fourth barrier lies in the division between policymaking and policy implementation. Policymaking (design and selection of policy instruments) is often shaped as an interactive process, whereby different actors with different interests are involved. Policy implementation, however, is mostly seen as a rather technical process that does not require the continuation of the interactive process. This is also reflected in the kind of monitoring and evaluation in these processes, which do not tend to be interactive and participatory. This strong division between policymaking and implementation is, in most cases, also reflected in the involved governments' organization. Often, different departments are responsible for policymaking and policy implementation.

A fifth barrier can be described as the underestimation of the implementation phase. Very often a lot of (political) attention of governments and other stakeholders is devoted to the decision-making phase because of the perception that implementation is only a last step that can 'handle itself'. The project partners in that stage are already starting a new project, or other business requires their attention. The long-term span needed for flexible arrangements, however, is difficult to sustain.

The DHZ case – survey and game

Besides analysing barriers to the implementation of flexible arrangements, we aimed to design instruments for making policy arrangements more flexible. This resulted in a survey and a serious game for policymakers involved in the design of policy implementation for climate adaptation. These instruments were applied in the DHZ case.

The survey was used once, at the beginning of our involvement in this case. All the individual members of the project team filled in the survey individually. The most remarkable outcomes were: (1) almost everybody defined the main object of the project in terms of content and not in terms of governance and (2) almost unanimously, the 14 project members indicated that they thought it important that the implementation arrangement would be flexible. This would imply, for example, that agreements between stakeholders could be altered along the way and some space would be left for stakeholders to implement measures in their own way.

A reflection on these results led to the collective insight that more focus was needed on the governance of implementation of the designed adaptation measures. Besides that, the survey revealed a paradox between the project members' strong wish to design flexible arrangements on the one hand, and the need to design effective and accountable measures on the other. Clearly, the focus in the project documents at that time was much more on effectiveness and accountability than on flexibility and reflectiveness.

After this awareness was raised by the survey, the serious game was used to help to design more flexible arrangements in practice. This resulted in the following insights concerning the emergence of flexible arrangements within DHZ.

The first insight was that the stakeholders were rather reluctant to alter existing practices. Especially in the first round of the game where small-scale water management measures were proposed, the participants assumed that government organizations would 'naturally' take responsibility and provide a financial compensation. However, another strategy could have been to leave it to a private initiative – for example, stewardship. The participants also showed an implicit preference for certain – mostly well-known – policy instruments. For example, during the game, the participants would often refer to subsidies/compensation as these are often applied in current Dutch policymaking. Other policy instruments such as auctions, taxes, or legislation were often not even mentioned, or quickly set aside as not (politically) feasible.

In general, we could claim that the perceived wishes of politicians played an important role in the proposed design of policy arrangements and the choices of policy instruments. One problem was that these politicians were not involved in the project team that we facilitated. Another problem was that not all the relevant politicians were convinced of the underlying problem (drought as a result of climate change). This probably contributed to the participants' tendency to present the problems in more 'hard' or 'structured' terms and to avoid more 'soft' terms such as learning and uncertainty. Policy instruments with uncertain outcomes, based on learning and reflectiveness, do not really fit well in such an environment. Besides, some of the stakeholders still have strong ties with the agricultural community (e.g. water board and agrarian unions). This means that policy instruments that would involve more costs or less freedom for agrarians were not very popular either.

The game, however, did lead to more awareness on the part of the involved stakeholders about how they selected instruments and constructed policy arrangements for policy implementation in practice. They became more aware that the selection of policy instruments had become an implicit activity based on habit rather than a conscious weighing of pros and cons. The interventions presented them, in the words of one of the participants: 'with new ways of approaching problems and offered them a wider toolkit that helped policymakers to think outside the box'. In one session, this led to a more flexible line of reasoning. Initially, the participants chose subsidies as the most fitting instrument for policy implementation but, after a while, they added other instruments to this choice. The line of reasoning now became: if after a while subsidies failed to reach the desired results, we could try to apply other instruments. Eventually, the game also contributed to the introduction of instruments such as service level agreements in the DHZ implementation plan.

We can conclude that the game helped the stakeholders to think outside the box and beyond their well-known policy instruments. This did not always lead directly to radically altered implementation arrangements. It did, however, lead to the consideration of other policy instruments and combinations of policy instruments. This could be the first step to more flexible policy implementation in practice.

Empirical lessons for the use of action research

In our project, we combined case studies with action research. We derived two important insights from this approach. First of all, the practitioners very much welcomed the evidence-based nature of the case studies' results. Because implementation of climate adaptation is not yet common practice, the stakeholders were very much looking for experiences from other projects.

A second important insight was that there was a possible weakness in our sequential approach and the fact that the researchers from our team that worked on the cases were less involved in the action research process. This could perhaps have been prevented by creating more synergy and interaction within the research team concerning the outcome of the cases and the action research, for example by positioning the serious game – even more – as a boundary object (Carlile, 2002) between science and practice. Another solution would be to use one of the case study researchers as a liaison on the action research team.

A challenge in doing action research in practice is the dependence on stakeholder participation. In our experience, crucial factors include the timing of action research interventions and (often related) the urgency and amount of attention that practitioners are able to devote to it. In our case, we can conclude in hindsight that we intervened too early. DHZ was still working on research towards formulating strategies and simply did not have the time to look too far ahead. In a specific pilot within DHZ that we used to test the game, we were too late, as they were just on the brink of finishing. They perceived the action research process – and perhaps rightly so – as a complicating or time-consuming factor. For future action research, it is really important to pay a lot of attention to the start time. The same is true for interventions. Action researchers should have close connections and insights into daily practices. A solution could be to include one of the practitioners on the action research team. This could lead to more policy–science circulation.

In relation to connecting the action research method with practice, our research shows that it is very important that the research is accepted by the stakeholders involved. Within DHZ, the action research project was accepted mainly by a minority of stakeholders. These were the stakeholders that also had some kind of affinity with science and policy, and implementation and innovation processes, because of either their background or their practical experience. These were also the contact persons for our action research. The other stakeholders did participate in the action research, but were not really involved and embedded in the research. It might have been better to cover a longer time span with our action research in order to increase stakeholder involvement.

Conclusions

We started this chapter with two questions:

1   What are the major (main) barriers to the development of flexible arrangements in order to be able to deal with the implementation of climate adaptation measures?
2   In what way can action research add to the insights on these barriers and can it help to overcome these barriers?

The nine case studies on the implementation of adaptation measures revealed a lack of focus on flexible arrangements. Most of the cases even designed the implementation phase rather inflexibly and with a high level of detail. This implies the existence of powerful barriers that prevent these implementation arrangements from being designed in more flexible ways. The nine case studies revealed some of these barriers. The most important were:

• policymakers' preference for robust adaptation measures;
• the existence of detailed contracts without explicit opportunities for reflection and evaluation;
• the existence of a powerful actor that can enforce its preferred solutions on others;
• the division between policymaking and policy implementation (in the policy process and within the governmental bureaucracy); and
• the underestimation by all stakeholders of the implementation time duration.

In addition to these case study-based insights, the results of our action research showed that technical aspects of adaptation measures received much more attention than the design of the flexible implementation arrangements. The survey showed, for example, that most aspects important for the design of flexible arrangements had not yet been addressed. This can be partly explained by the phase in which DHZ was, but a better explanation might derive from stakeholders' persistent idea that, once the measures have been designed, implementation is just a small step. We could claim that implementation is still framed as a technical process of implementing well-described policy solutions for well-known policy problems. Uncertainty about the effects of climate change, about circumstances, and about the effectiveness of particular policy instruments is preferably reduced or ignored. This would imply a combination of lack of knowledge (about, for example, adaptive governance, policy instruments, flexible arrangements) and the way policymakers deal with uncertainty (reducing it rather than dealing with it).

The former was also visible in the way in which the underlying policy problem, that is, drought, was formulated within DHZ: the most extreme climate change scenario was selected. The possible drought effects and the selection of adaptation measures, including the calculations of costs, were based on this particular scenario. This way of dealing with uncertainty also reduces the need to be flexible in policy and implementation. In our action research, we did not address this issue actively as this choice had already been made.

Concerning the second research question, we can conclude that most of the barriers found in the case studies can also been identified in the action research case. Furthermore, action research, although primarily reflective in nature in our case, has revealed some additional insights. One such insight is the reluctance of stakeholders to introduce flexible arrangements in the process. This can probably be explained by a lack of knowledge and information about these arrangements and about adaptive governance more broadly. Another new element was the politicians' role. The actors on the administrative level anticipated the perceived wishes of their politicians to reduce uncertainty in the policy process. This was done by structuring both the policy problem and the available solutions, and thus reducing flexibility.

Reflecting on our research methods, we have combined traditional research methods (case studies and desk research) with action research (action science). These methods have complemented each other and were both very helpful in identifying barriers to flexible policy arrangements. In the case studies and desk research, the information was limited, deduced, and based only on a selection of documents and stakeholder interviews. The action research with DHZ was based only on one case, but the great strength of the study is the opportunity to experience and gain in-depth insights into the stakeholder process and motives of individual actors. However, this exercise also reveals the problem of translating scientific knowledge (even based on empirical research) into practical guidelines for policymakers. These policymakers do in fact have to deal with barriers other than those defined by researchers (see van Buuren *et al.*, Chapter 1, this volume).

Obviously, there are multiple ways to do this. In our research, we tried to translate the scientific knowledge on flexible arrangements and barriers into a survey and a serious game. The serious game was played in an interdisciplinary way, with scientists and the local stakeholders together. This proved to be a challenging approach that required a high level of flexibility (and creativity) on the part of the researchers involved. This included, in some cases, moving along the edges of scientifically sound statements.

Although originally intended to do so, the action research in the DHZ case did not specifically lead to the design of more flexible policy arrangements in the implementation of this adaptation plan. It did, however, lead to more awareness, and thus action research did enhance reflexivity among the policymakers involved on how to select instruments and construct policy arrangements for policy implementation in practice. They became more aware that the selection of policy instruments had become an implicit activity that was based on habit rather than a conscious weighing up of pros and cons. The interventions presented them with new ways of approaching problems and offered them with a wider toolkit that helped policymakers to think out of the box. Eventually, this contributed to the introduction of instruments such as service level agreements in the DHZ implementation plan. So, the action research really led to cognitive awareness of the possible flexible arrangements for climate adaptation.

To conclude this chapter about using action research, we did create awareness and enriched insights into questions that to date have only been answered theoretically or by case study. This was also stated by the stakeholders who participated in either the survey or the game. Action research offers the opportunity to get detailed and embedded insights into actors' motivations concerning the design of flexible arrangements. Action research, more than other methods, pays attention to the contextual aspects of choosing arrangements. Finally, action research helps to rethink in-depth the barriers with which actors are confronted in climate adaptation strategies.

References

Abt, C.C. (1975) *Serious games*, New York: Viking.

Adger, N.W., N.W. Arnell and E.L. Tompkins (2005) Successful adaptation to climate change across scales, *Global Environmental Change*, 15(2): 77–86.

Amundsen, H., F. Berglund and H. Westdkogô (2010) Overcoming barriers to climate change adaptation: a question of multilevel governance? *Environment and Planning C: Government and Policy*, 28(2): 276–289.

Argyris, C. (1996) Actionable knowledge: design causality in the service of consequential theory, *The Journal of Applied Behavioral Science*, 32(4): 390–406.

Armstrong, J.S. (1979) The natural learning project, *Journal of Experimental Learning and Simulation*, 1(1): 5–12.

Bai, X. (2007) Integrating global environmental concerns into urban management: the scale and readiness arguments, *Journal of Industrial Ecology*, 11(2): 15–29.

Bekebrede, G. (2010) *Experiencing complexity: a gaming approach for understanding infrastructure systems*, Delft: Next Generation Infrastructures Foundation.

Biesbroek, R., J. Klostermann, C. Termeer and P. Kabat (2011) Barriers to climate change adaptation in the Netherlands, *Climate Law*, 2(2): 181–199.

Bradach, J.L. and R.G. Eccles (1989) Price, authority and trust: from ideal types to plural forms, *Annual Review of Sociology*, 15: 97–118.

Buuren, M.W. van, P.P.J. Driessen, H.J.F.M. van Rijswick and G.R. Teisman (2014) Towards legitimate governance strategies for climate adaptation: Combining insights from legal, planning and democratic perspectives, *Regional Environmental Change*, 14(3): 1021–1033.

Carlile, P.R. (2002) A pragmatic view of knowledge and boundaries: boundary objects in new product development, *Organization Science*, 13(4): 442–455.

Cooney, R. and A.T. Lang (2007) Taking uncertainty seriously: adaptive governance and international trade, *European Journal of International Law*, 18(3): 523–551.

Crabtree, B. and W. Miller (1999) *Doing qualitative research*, 2nd edn, Thousand Oaks, CA: Sage

Crosby, B.L. (1996) Policy implementation: the organizational challenge, *World Development*, 24(9): 1403–1415.

Cundill, G. and C. Fabricius (2009) Monitoring in adaptive co-management: toward a learning based approach, *Journal of Environmental Management*, 90(11): 3205–3211.

DHZ (2014) *Op weg naar een strategie en uitvoeringsprogramma voor de regio's oost en zuid: sparen, aanvoeren, accepteren en adapteren*, 's-Hertogenbosch: Province of North Brabant.

Duke, R.D. and J. Geurts (2004) *Policy games for strategic management: pathways into the unknown*, Amsterdam: Dutch University Press.

Dupuis, J. and P. Knoepfel (2013) The adaptation policy paradox: the implementation deficit of policies framed as climate change adaptation, *Ecology and Society*, 18(4): 31.

Edelenbos, J. and J. Eshuis (2011) The interplay between trust and control in governance processes: a conceptual and empirical investigation, *Administration and Society*, 44(6): 647–674.

Edelenbos, J., A. van Buuren and E.H. Klijn (2013) Connective capacities of network managers: a comparative study of management styles in eight regional governance networks, *Public Management Review*, 15(1): 131–159.

Ellen, G.J., B. Breman, J.J. Dijk, R.J.M. Franssen, A.M. Keessen, W. Kuindersma, F. van Lamoen, M.W. van Buuren, C.J.W.G. van Leeuwen and D. van Soest (2014) *De implementatie van adaptatie: barrières en mogelijkheden voor flexibele governance-arrangementen*, KvK report 114/2014, Utrecht: Kennis voor Klimaat.

Flyvbjerg, B. (2006) Five misunderstandings about case-study research, *Qualitative Inquiry*, 12(2): 219–245.

Folke, C., T. Hahn, P. Olsson and J. Norberg (2005) Adaptive governance of social-ecological systems, *Annual Review of Environment and Resources*, 30: 441–473.

Gambetta, D. (1988) *Trust: making and breaking cooperative relations*, New York: Basil Blackwell.

Gibbs, G. (1987) *Learning by doing*, London: FEU.

Giddens, A. (2009) *The politics of climate change*, Cambridge: Polity Press.

Gunderson, L. (1999) Resilience, flexibility and adaptive management: antidotes for spurious certitude? *Ecology and Society*, 3(1): 7.

Gupta, J., C.J.A.M. Termeer, E. Bergsma, G.R. Biesbroek, M.A. van den Brink, P. Jong, J.E.M. Klostermann, S.V. Meijerink and S.G. Nooteboom (2011) *Institutions for adaptation: do institutions allow society to adapt to the impacts of climate change?* KvR report 039/11, Nieuwegein: Klimaat voor Ruimte.

Haasnoot, M., J.H. Kwakkel, W.E. Walker and J. ter Maat (2013) Dynamic adaptive policy pathways: a method for crafting robust decisions for a deeply uncertain world, *Global Environmental Change*, 23(2): 485–498.

Hallegatte, S. (2009) Strategies to adapt to an uncertain climate change, *Global Environmental Change*, 19(2): 240–247.

Harteveld, C. (2010) *Triadic game design*, Heidelberg: Springer.

Hovi, J., D.F. Sprinz and A. Underdal (2009) Implementing long-term climate policy: time inconsistency, domestic politics, international anarchy, *Global Environmental Politics*, 9(3): 20–39.

Kolb, D.A. (1984) *Experiential learning: experience as the source of learning and development*, Upper Saddle River, NJ: Prentice-Hall.

Lather, P. (1992) Critical frames in educational research: feminist and post-structural perspectives, *Theory into Practice*, 31(2): 87–99.

Lee, K.N. (1999) Appraising adaptive management, *Conservation Ecology*, 3(2): 3.

Majone, G. and A. Wildavsky (1978) Implementation as evolution, in H.E. Freeman (ed.) *Policy studies annual review: Volume 2*, 103–117, Beverly Hills, CA: Sage.

Mayer, I.S. (2009) The gaming of policy and the politics of gaming: a review, *Simulation & Gaming*, 40(6): 825–862.

Mess, H. L. P., J. Dijk, D. van Soest, P. P. J. Driessen, M. H. F. M. W vas Rijswick and H. Runhaar (2014) A method for the deliberate and deliberative selection of policy instrument mixes for climate hange adaptation., *Ecology and Society*, 19(2): 58

Nicol, D.J. and D. Macfarlane-Dick (2006) Formative assessment and self-regulated learning: a model and seven principles of good feedback practice, *Studies in Higher Education*, 31(2): 199–218.

Olsson, P., L.H. Gunderson, S.R. Carpenter, P. Ryan, L. Lebel, C. Folke and C.S. Holling (2006) Shooting the rapids: navigating transitions to adaptive governance of social-ecological systems, *Ecology and Society*, 11(1): 18.

Pahl-Wostl, C. (2007) Transitions towards adaptive management of water facing climate and global change, *Water Resources Management*, 21(1): 49–62.

Plummer, R. and D. Armitage (2007) A resilience-based framework for evaluating adaptive co-management: linking ecology, economics and society in a complex world, *Ecological Economics*, 61(1): 62–74.

Plummer, R., D.R. Armitage and R.C. de Loë (2013) Adaptive comanagement and its relationship to environmental governance, *Ecology and Society*, 18(1): 21.

Reason, P. (2003) Three approaches to participative inquiry, in N.K. Denzin and Y.S. Lincoln (eds) *Strategies of qualitative inquiry*, 261–291, Thousand Oaks, CA: Sage.

Ring, P.A. and A. van de Ven (1994) Developmental processes of cooperative inter-organizational relationships, *Academy of Management Review*, 19(1): 90–118.

Robottom, I.M. and E.P. Hart (1993) *Research in environmental education: engaging the debate*, Geelong: Deakin University Press.

Royal Haskoning (2012) *Regionale knelpuntenanalyse Zuid-Nederland*. Report commissioned by Stuurgroep Deltaplan Hoge Zandgronden. 's Hertogenbosch: Royal Haskoning.

Sabatier, P. and D. Mazmanian (1980) The implementation of public policy: a framework of analysis, *Policy Studies Journal*, 8(4): 538–560.

Termeer, C.J., A. Dewulf and M. van Lieshout (2010) Disentangling scale approaches in governance research: comparing monocentric, multilevel, and adaptive governance, *Ecology & Society*, 15(4): 29.

Thompson, M., R. Ellis, and A. Wildavsky (1990) *Cultural Theory*, Boulder, CO: Westview.

Uittenbroek, C.J., L.B. Janssen-Jansen and H.A.C. Runhaar (2012) Mainstreaming climate adaptation into urban planning: overcoming barriers, seizing opportunities and evaluating the results in two Dutch case studies, *Regional Environmental Change*, 13(2): 399–411.

Underdal, A. (2010) Complexity and challenges of long-term environmental governance, *Global Environmental Change*, 20(3): 386–393.

Wallington, T.J. and S.A. Moore (2005) Ecology, values and objectivity: progressing the debate, *BioScience*, 55(10): 873–878.

Warner, J.F., J. Warner, A. van Buuren and J. Edelenbos (eds) (2013) *Making space for the river: governance experiences with multifunctional river flood management in the US and Europe*, London: IWA Publishing.

Yin, R.K. (2003) *Case study research: design and methods*, 3rd edn, Thousand Oaks, CA: Sage.

8 Scaling innovation in climate change planning
Serious gaming in Portland and Copenhagen

Patrick Driscoll and Martin Lehmann

Introduction

For over 20 years, cities around the globe have been developing and implementing planning strategies to address climate change by primarily focusing on mitigation, but in the last few years adaptation has risen rapidly as a strategic priority all over the world (Carmin *et al.*, 2012; Biesbroek *et al.*, 2009). Modern municipal planning authorities have come under increasing pressure to innovate and respond to a growing demand from citizens and politicians alike to respond effectively to the climate change challenge, a challenge that includes balancing the synergies, conflicts, and trade-offs between mitigation, adaptation, and sustainable urban development. A number of complex governance issues are present within this field, including: unclear legal and regulatory environments (Carmin *et al.*, 2012), new types of cooperative relationships between municipal governments, utilities, the finance and insurance sectors, citizens, non-governmental organizations, and private businesses (Betsill and Bulkeley, 2007; Juhola and Westerhoff, 2011; Rosenzweig *et al.*, 2011), conflicting risk perceptions (moral hazard), the science–policy interface in a new arena of planning (Kahan *et al.*, 2012), and negotiating trade-offs between mitigation, adaptation, and sustainable development goals (Swart and Raes, 2007; Tol, 2005; Wilbanks and Sathaye, 2007).

Our theoretical understanding of the modern municipal planning organization is informed by Røvik's (1998) conception of a multi-standard organization that possesses a high potential for acquisition, assimilation, and exploitation of new knowledge and practices. The guiding research question is the extent to which serious games can stimulate reflection and learning and thereby help develop a deeper understanding of the decision basis for planners.

This chapter presents results derived from an ongoing comparative case study of Copenhagen and Portland investigating the decision-making practices of planners working with mitigation and adaptation, where serious games comprise part of the data collection methodology. A growing body of evidence suggests that serious games can have a positive effect on communication and learning, both in the short term and over the long term (see, for example, Haug *et al.*, 2011; Juhola *et al.*, 2013; Meijer, 2010; Patt *et al.*, 2009; Rijcken *et al.*, 2012).

The contemporary concept of serious games can be traced back to Abt (1970), based upon his experiences with war planning, and how games could be used to combine play, education, and learning. Serious games have been used within climate change planning since at least 1983 (Ulrich, 1997). In the 20 years since then, Reckien and Eisenack (2013) found over 50 serious games developed to deal with either the mitigation or the adaptation aspects of climate change. Academics were early drivers of the trend, but more recently governmental/non-governmental organizations and private businesses are developing a majority of serious climate change games. Well over 80 per cent of the games surveyed in the study mainly addressed mitigation, 40 per cent addressed some aspect of adaptation, and fewer than 10 per cent examined development issues, but none of the games incorporated all three of these aspects within the game space (Reckien and Eisenack, 2013).

Two notable methodological innovations are presented here that may be of use to action research and the governance of climate change. One is the combination of mitigation, adaptation, and sustainable urban development within the same gaming space; the other is the development and use of social strategy games as a data collection tool, combining digital video, participant observation, and interviews. 'Broken Cities' was originally developed for the Nordic Climate Fair held in Helsinki, 2010, in collaboration with Aalto University, the International Red Cross/Red Crescent Climate Centre, Parsons The New School for Design, and Aalborg University. The early goal of the game was to communicate to planning students the complexities of the trade-offs between mitigation, adaptation, and sustainable urban development (see Juhola *et al.*, 2013 for a more detailed description of the game structure and mechanics). The game served its original purpose well, but it also became quickly apparent that it could be quite useful as a data collection device, particularly for gaining insight into the decision-making space of planners as well as being a potential stimulus for learning.

This chapter first presents the governance cases of climate change planning in Copenhagen and Portland, including a brief overview of the mitigation and adaptation strategies, the political, legal, and regulatory context, and a review of the synergies, conflicts, and trade-offs present within the municipal strategies. Thereafter, the methods and materials employed during the research, including a detailed description of the mechanics of 'Broken Cities', the connection to the comparative cases, and the use of serious games within action research, are described. Next, the empirical results from the use of 'Broken Cities' are presented along with a discussion of some of the promises and pitfalls of an action research approach in the field of climate adaptation governance. The chapter ends with some reflections on the use of serious games in action research and how our experiences can be used to realize their potential more effectively.

Governance cases

The Region of Greater Copenhagen encompasses the cities of Copenhagen (pop. 570,000), Frederiksberg (pop. 103,000), and 28 surrounding municipalities with almost 1,750,000 inhabitants (Statistics Denmark, 2013). The total population of the functional urban region on the Danish side of the Øresund bridge (including the island of Zealand) is approximately 2.4 million, accounting for 43 per cent of the entire Danish population (Oresundregion, 2013). Post-war urban development in the region has been guided to a large extent by the 1947 Finger Plan, which established five development corridors concentrated around the trunk lines of the regional train network (S-tog). The most recent revision of the Finger Plan in 2007 reinforces and extends these principles of concentrating development along public transport corridors (Miljøministeret, 2007). Although development patterns since the end of World War II have resulted in extensive expansions of urbanized areas outside the Finger Plan, the city of Copenhagen still possesses considerable residential and workplace location appeal (Næss, 2005), and there is a growing repopulation of the central urban core (Næss, 2011).

The consolidated Portland metropolitan area spans seven counties in two states (Oregon and Washington) with a combined population of over 2.2 million (United States Census Bureau, 2010). The city of Portland (pop. 600,000) is the largest in Oregon and is a central hub of service, manufacturing, and logistics in the region. Portland's planning system is well known for its embrace of urban growth controls, densification efforts, and transit-oriented development policies that attempt to closely link land use, development and transportation planning decisions (Abbott & Margheim, 2008).

Copenhagen and Portland political and planning context

The political structure of Copenhagen Municipality consists of a city council of 55 members, including one lord mayor and six vice-mayors, with each of the vice-mayors given leadership of a permanent standing committee. The climate change planning functions reside within the Technical and Environmental Department, which is the second most politically powerful position in the city, second to the lord mayor. The Climate Unit within the city has 22 planners evenly split between mitigation and adaptation functions. The regional authority is relatively weak, leaving much of the climate change planning authority in the hands of the city and the state, split between the Ministry of Climate and Energy (mitigation) and the Ministry of the Environment (planning law and adaptation).

In Portland, there are five city commissioners, including the mayor, and the lead agency on both the mitigation and adaptation strategies is the Bureau of Planning and Sustainability (BPS), which works in close cooperation with Multnomah County and other city agencies, such as the Bureau of Environmental Services, the Portland Bureau of Transportation, and the Multnomah County Health Department. There are approximately 10 planners working with climate change at the City of Portland, only two of whom work part-time on adaptation. An additional governmental layer is present in Portland that does not exist in Copenhagen; that is, a democratically elected regional government with legal authority to control land use and transport planning in the Portland Metropolitan Area. The State of Oregon provides some resources and guidance on mitigation and adaptation planning.

Climate change mitigation and adaptation planning in Copenhagen and Portland

For much of the 1990s and 2000s, both cities have devoted significant planning attention to lowering greenhouse gas (GHG) emissions, and consequently very little to climate adaptation strategies – until 2011 with the release of the Copenhagen Climate Adaptation Strategy and 2014 with the release of the Portland Climate Change Prevention Strategy. After three intense storm events that led to widespread flooding in 2010 and 2011, the City of Copenhagen adopted an adaptation strategy in 2011 (City of Copenhagen, 2011) and subsequently, in 2012, a Cloudburst Strategy for eight of the most vulnerable areas within the city to develop resilient human, environmental, and infrastructure systems (City of Copenhagen, 2012a).

The Municipality of Copenhagen has had a long-standing policy of addressing climate change, primarily through mitigation measures embedded in Local Agenda 21 plans in the 1990s and 2000s, and most recently in the Climate Plan adopted in 2009 (City of Copenhagen, 2009). Copenhagen has adopted a goal of carbon neutrality by 2025, comprising actual reductions of annual carbon emissions from 1.9 million tons of CO_2 per annum in 2011 to 1.16 million tons by 2025, combined with offsets from renewable power generation (coal to wind) (City of Copenhagen, 2012b).

In 1993, Portland was the first major city in the United States to adopt a carbon dioxide mitigation strategy; Multnomah County joined Portland to create a shared city/county strategy in 2001 (City of Portland and Multnomah County, 2001). To date, per capita greenhouse emissions in Multnomah County have declined by 1 per cent, whereas the trend line for the US as a whole has seen a 13 per cent increase in emissions (from a 1990 baseline) (City of Portland and Multnomah County, 2009, 2012). The most recent joint Climate Action Plan adopted by both the City of Portland and Multnomah County in 2009 (the new 2013 plan is delayed and will not be released until late 2014) recognizes that much more radical reductions in future greenhouse gas emissions will be necessary. The city and county have committed themselves to a goal of 40 per cent reduction in GHG emissions by 2030 and an 80 per cent reduction by 2050 (City of Portland and Multnomah County, 2009).

To date, there have been only halting efforts to plan for or capture synergies between mitigation and adaptation strategies and the respective cities' sustainable urban development goals (see Tables 8.1 and 8.2 for more detailed, city-specific assessments[1]). For example, the Copenhagen adaptation strategy identifies green roofs and façades as a desirable planning goal for, among other reasons, reductions in the urban heat island effect, better air quality, better insulation, lower energy usage, a larger area for wildlife, and better stormwater management. But these effects also serve to lower GHG emissions by reducing energy consumption, obviating the need for large expansions of the sewer system and flood control measures, reducing or eliminating the need for air conditioning, and reducing the reliance on traditional, energy-intensive roofing materials. The failure to integrate these types of mitigation and adaptation efforts is a missed opportunity for at least two reasons. First, cities will need to build support for far more radical reductions in GHG emissions in the near future, and a relatively easy way to garner that support is to focus attention on policies that can deliver multiple co-benefits. Second, there is a significant risk of implementing maladaptive strategies that may decrease GHG emissions in the short run but lead to higher emissions over the long run if active interventions are necessary to sustain them.

One point that should be stressed is that, even in situations where mitigation and adaptation strategies have synergies between them, they may still conflict with other sustainable development goals such as equity or environmental protection. To give a more detailed example of the types of trade-offs present in Portland, the 20-minute neighbourhood goal (whereby non-work trips should be doable by foot or bicycle within 20 minutes) has the tendency, all other things being equal, to reinforce and extend existing inequalities between the more compact (richer and whiter) inner-city neighbourhoods with the lower density (poorer and non-white) outer neighbourhoods that are fast becoming the new ghettos in Portland. Rising energy prices and increased residential building efficiencies may also have a disproportionate impact on poorer residents who tend to live in sub-standard rental housing far from accessible public transportation, thereby necessitating longer, car-based trips.

Moreover, the spatial and urban planning regimes in both cities have been poorly served by the perceived distinctions between mitigation and adaptation. Treated separately, there is a danger that adaptation policies will trigger increases in GHG emissions (maladaptation) and that mitigating policies will trigger increases in societal vulnerability to climate change (undermining sustainable development goals). An example of this can be seen in Figure 8.1, which describes how the City of Portland sees the interface between mitigation and adaptation. One of the main adaptation challenges for Portland is urban heat island effects, which are exacerbated by income inequalities in the city. Through policies of reducing car travel and increasing energy efficiency, which have the effect of raising housing prices and pushing poorer people further away from jobs, public transport, and daily services, the impacts of poorly designed climate change policies fall disproportionately harder on weaker segments of the population, exacerbating an already unequal distribution.

Furthermore, researching climate change mitigation and adaptation separately may reinforce this unfortunate situation (Cohen and Waddell, 2009), leading to avoidable conflicts between mitigation and adaptation measures and strategies and obscuring the necessary trade-offs between mitigation, adaptation, and sustainable development. One specific example from Copenhagen (Table 8.2) is that the strategy to handle increasing rainwater loads in the city is being financed by a high sewer charge that is not proportionate to income, thereby hitting poorer residents much harder than richer ones, even though many of the improvements derived from the adaptation strategies are likely to accrue to the wealthiest residents living in the compact urban core of the city.

Table 8.1 Selected synergies, conflicts, and trade-offs between climate change goals in Portland

Mitigation objective	Primary planning sector	Secondary planning sector	Adaptation objective	Adaptation action	Synergy or conflict	Trade-offs
20-minute complete neighbourhood concept (e.g. most non-work trips should be within 20 minutes travel time by foot or bicycle)	Built environment	Transport Energy	Successful adaptation	Green infrastructure	Potential conflict due to space requirements	Mal-distributed infrastructure and opportunities; Residence and employment location freedom; Bias toward centre city; More vulnerable critical infrastructure
Reduce daily vehicle miles travelled (VMT) by 30%	Transport	Urban morphology Energy	N/A	N/A	N/A	Potential harm to poorer citizens in areas poorly served by transit; Loss of access to employment; Lower mobility leading to lower economic growth rates
Reduce energy usage of pre-2010 building stock by 25%	Built environment, Energy	Urban morphology	Buildings that can adapt to changing climates	Influence state-wide building codes	Synergy	Higher up-front housing costs; Occupancy behaviour of buildings restricted
10% of energy produced by on-site renewable sources	Built environment Energy	Urban morphology	Building that can adapt to changing climates	Influence state-wide building codes	Potential synergy Potential conflict (for example densification efforts and solar rights)	Denser developments; Solar rights conflict; Land use/siting conflicts over wind or solar
Zero net GHG emissions from new homes	Built environment Energy	Urban morphology	Adaptive buildings	Alter building codes	Synergy	Higher up-front building costs, putting pressure on low-cost housing

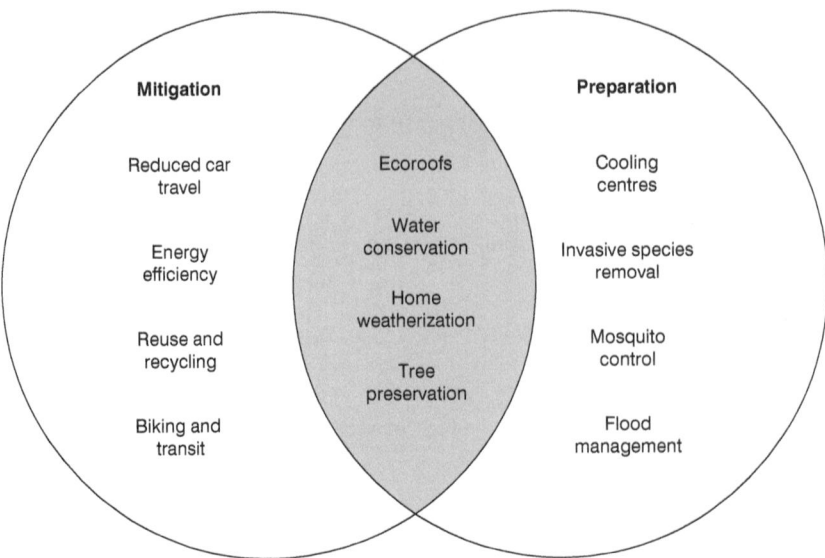

Figure 8.1 Overlaps between mitigation and adaptation in Portland (source: City of Portland and Multnomah County, 2014)

These kinds of trade-offs identified here between urban development, compact cities, environmental protection, social equity, mobility, and economic development are not entirely new. What is different about the relationship between climate change mitigation and adaptation strategies and other environmental goals is that successfully creating a low emission profile for present and future development will require a re-examination of the underlying development pathway, where resource efficiency gains are frequently consumed by increased levels of affluence and consumption.

There are two main points to highlight here. First, climate change planning is still mostly rooted within an energy and environmental framework. This is an understandable position relative to mitigation efforts, given that a majority of GHG emissions are connected to the energy generation and transport sectors, but is strikingly less so when applied to adaptation policies and plans. If cities and regions are serious about lowering their GHG emission profiles and building resiliency into the urban fabric, then planners will need to work towards integrating climate change planning more firmly within the existing planning framework, rather than treating climate change as a bolted-on appendage.

Second, as adaptation strategies are frequently developed and implemented independent of mitigation strategies, the lack of a strategic framework to analyse synergies, conflicts, and trade-offs between mitigation and adaptation goals will likely lead to missed opportunities and create avoidable policy conflicts. As noted above, many of the climate change goals articulated by both the Portland and Copenhagen planning authorities do have synergistic elements with other planning goals, such as social inclusion, green space provision, safer streets, less

Table 8.2 Selected synergies, conflicts, and trade-offs between climate change goals in Copenhagen

Mitigation objective	Primary planning sector	Secondary planning sector	Adaptation objective	Adaptation action	Synergy or conflict	Trade-offs
Renewable energy supply for combined heat/power plants	Energy	Urban morphology Built environment	Protection of critical infrastructure from storm surges and extreme rainfall	Raise dykes Expand sewer system capacity Moving vulnerable infrastructure	Potential conflict due to reliance upon high energy systems to protect existing centralized energy infrastructure	Institutional conflicts Political and economic risk Land use/siting conflicts over wind or solar energy Greater vulnerability for critical infrastructure
Increase cycling to 50% mode share	Transport	Urban morphology Energy	Use existing infrastructure to retain and divert storm water	Use of green streets to channel rainwater	Synergy	None identified
Reduce energy usage of building stock by 7.5%	Built environment Energy	Urban morphology	Expand green roofs and façades	Demonstration projects in Copenhagen Disconnection of rainwater from the sewer system	Synergy	Opposition from existing district heating provider Increased housing costs
Traffic re-routing, lane reduction measures, and parking restrictions	Transport	Urban morphology Built environment	Pocket parks and permeable surfaces to locally divert rainwater	Construct new parks and integrate green infrastructure into new developments	Synergy	Potential harm to poorer citizens in areas poorly served by transit Loss of access to employment Lower mobility leading to lower economic growth rates

car traffic, and more compact urban development patterns. There are, however, significant conflicts between, for example, space-hungry adaptation strategies designed to handle storm water runoff and emission-lowering goals of compact urban development and increased densities within built-up areas.

Methods and materials

Following the typology of approaches to action research presented in Chapter 2 by Huntjens *et al.*, the present study is informed by an action science/inquiry approach rooted within a pragmatic philosophy of science (Greenwood and Levin, 1998; Peirce, 1982; Reason, 2003) (see Table 8.3), with a mid-level involvement of participants, whereby planners were consulted about the overall structure of the research as well as presented with the mid-way and final results for critical feedback. Some of the pitfalls encountered, which are elaborated in more detail in the sub-section on reflections, include a lack of engagement by the planners in the project, and the serious game utilized as a data collection tool could have used more experiential insight in the development and deployment of the game. This is partly due to the fact that, in the initial stages of the overall research design, action research was not an explicit component of the methodology but has become woven into the design and execution of the project.

In action science, the central perspective is the identification of theories and implicit cognitive models that actors use to guide their behaviour as well as the actual behaviour itself (Reason, 1994). Action inquiry, in addition to the aforementioned, also addresses outcomes and measures them empirically, and attends to the quality of the researcher's own attention. Finally, action inquiry attends to the issue of how organizations and communities can be transformed into collaborative, self-reflective communities of inquiry. With this in mind,

Table 8.3 Action research typology

Approach	Level of intensity of engagement	Level of involvement	Opportunities with this approach	Pitfalls with this approach
Action science – explicit linkages to pragmatic philosophy of science	Consultation – planners have been extensively briefed on the goals and outcomes of the research	Level 3 (reflection) – linked to the level of intensity	Ability to preserve critical distance between research objectives and planning objectives	Lack of engagement with planners, leading to low levels of ownership of the findings Missing out on the collective insights and experience when developing the serious game

introducing serious gaming is a way to engage the researchers' actions with others' actions in a self-reflective way so that all involved become more conscious about their behaviour and the underlying theories, purposes, and strategies behind this behaviour (Reason, 1994). In this way, a more traditional scientific framework of explanation and (possibly) prediction can be imposed; this is important when one is interested in the science of the action itself.

For both Copenhagen and Portland, the following types of planning strategies and other documents, including action plans, have been included in the data collection: climate mitigation plans, climate adaptation plans, long-range (longer than 10 years) master plans, short-range (between 5 and 10 years) municipal plans, land-use plans at both district and local level, transport plans, energy plans, environment plans, economic development strategies, and social equity strategies. These documents were used to build the broad analytic framework of the mitigation, adaptation, and sustainable urban development goals. The selected synergies, conflicts, and trade-offs presented in Tables 8.1 and 8.2 were chosen in the following manner. For mitigation objectives, we selected those measures designed to deliver the largest emission reductions (built environment, energy, and transport). For adaptation measures, we selected those afforded the highest near-term priorities, such as storm and flooding events, and urban heat island effects.

Two gaming sessions were organized in both Copenhagen and Portland, with a mix of 15 to 20 planners, students, and interested citizens lasting two to three hours each. The sessions were recorded with video cameras, and researchers were present to observe the game play, take notes, and de-brief the participants at the end of the gaming sessions. The use of 'Broken Cities' as a data collection tool within this project was an emergent phenomenon that had not been part of the original research design, but rather arose from the rich interactions observed among the planning and engineering students who were involved in a smaller pilot project conducted in 2010–2011 in the US, Finland, and Denmark.

One of the more remarkable elements of the game play was the nature of the double-loop learning that was happening, where the players demonstrated a high level of understanding of the complex linkages between their in-game decisions and real-world impacts of climate change. The game was refined and developed further by a team of researchers at Aalborg University, introducing new game elements (such as currency) and more formal rules (such as regulatory policies) to more closely mirror the real-world challenges faced by planners working with adaptation and mitigation in an urban space. Additionally, the game was printed in a box set and distributed to researchers, planners, and students for their own use and distributed online under a Creative Commons BY-NC-SA 3.0 licence.

Game design and mechanics

'Broken Cities' is designed for four or five (teams of) players, depending on the total numbers of players in a session – which has been up to 75 people at a time.

There are four developers, and the fifth player plays the role of local government to solve disputes, track emissions and rental income, and initiate policies in order to keep emissions under certain thresholds. Game play in 'Broken Cities'[2] is a turn-based urban development game, where each player, starting with the yellow quadrant, begins by collecting rental income and then making a series of investment choices (three different housing types, retrofit of existing buildings, green spaces, eco-parks, shopping malls, and so forth). Players may also opt to do nothing and pass to the next player. At the end of an individual player's turn, rental income and CO_2 emissions are calculated. Players must also keep track of cumulative atmospheric damage, roughly simulating the relationship between individual choices and global GHG concentrations. Players' decisions have spillover effects as well, both positive and negative.

For example, players can choose to buy green space or eco-parks, thereby lowering their individual emissions footprint and raising the rents on adjacent properties by making them more attractive and valuable. The placement of the green space or eco-parks may thus create positive externalities for other players, and consequently they may wish to share the costs and benefits. Conversely, players may choose to capture all of the benefits for themselves and ignore the collective impacts on the other players. The game ends when one or more players attain 50 units of rental income or the cumulative atmospheric damage reaches 42 (Figure 8.2).

All players are given currency with which to make purchases and cheat sheets, where information is provided about the costs, CO_2 emissions, rental income, and co-benefits of each of the development options. Players can adopt any number of different strategies, none of which is pre-determined by the game mechanics. Since the overall research aim is to investigate how strategy games can deepen understanding of complex issues and enhance the innovative capacity of planners, the designers intentionally created space for players to devise their own unique strategies in dialogue with the local and regional government representatives who are responsible for maintaining the emissions cap. Since it is not possible to arrive at a theoretical optimal mix between mitigation, adaptation, and sustainable development goals, the players must, through a combination of individual profit seeking and collective decision making, arrive at a set of strategies that lower overall carbon emissions, ensure the resiliency of the built environment, and address the needs of the 'population' of their city.

As the game progresses, carbon emissions typically rise in the early stages since most players are focused on earning rental income. The atmospheric damage calculator has fixed points; when the players reach these points, they must draw event cards. There are six types of impact cards, including hurricanes, droughts, disease vectors, landslides, coastal flooding, and a dummy card (an unmaterialized threat). The impacts are often distributed unevenly across the players' social and environmental strata. For example, the flood card destroys all low-cost housing next to water, but does not affect more expensive housing types, whereas landslides wipe out any structures adjacent to the forest areas on the board.

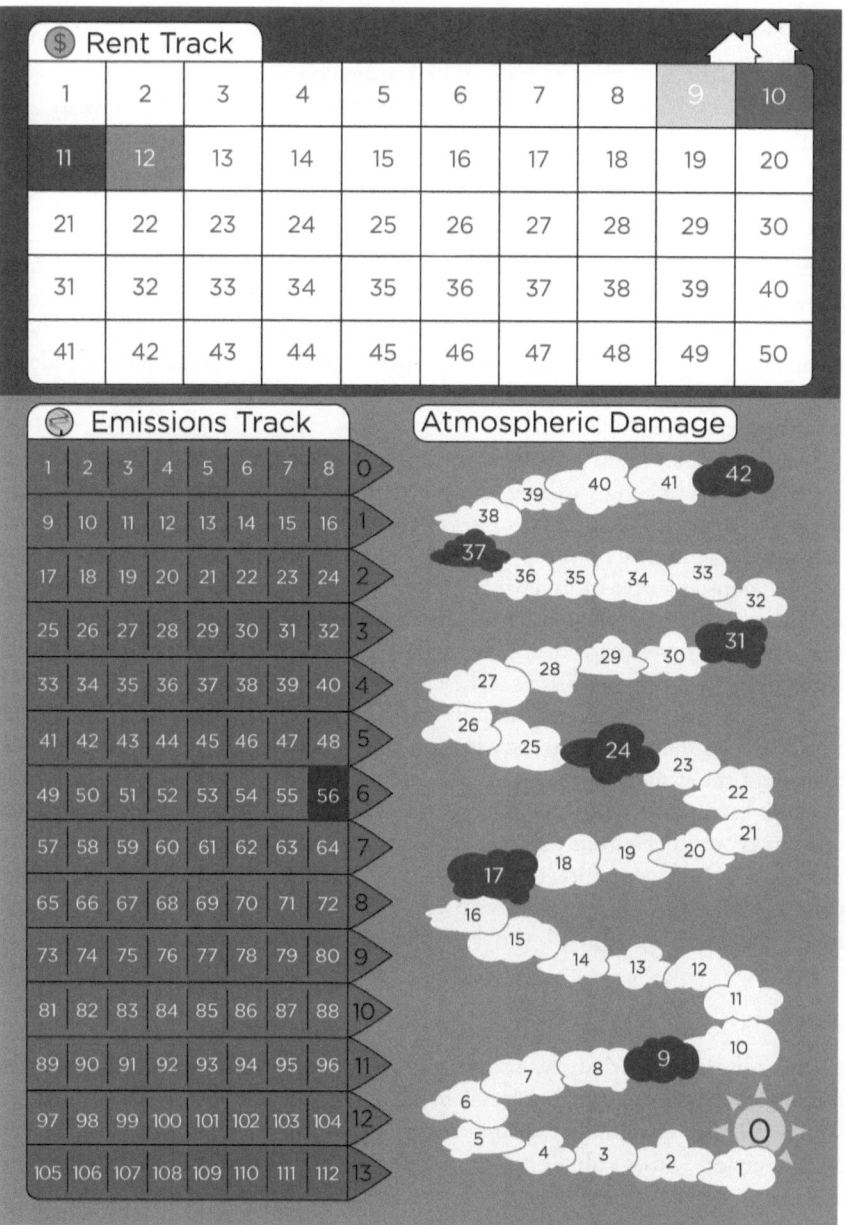

Figure 8.2 Rent, emissions, and atmospheric damage tracker

Empirical results

One of 'Broken Cities' more immediate impacts on planners in both cities was that they were compelled by the structure of the game to interact with one another in ways that they typically do not in the course of their normal working day. In Copenhagen, for example, the planners from the Parks and Nature unit played with counterparts from the Centre for Traffic and the Centre for Strategic Urban Development that they rarely, if ever, encountered. In Portland, the planners played with citizens and planning students from Portland State, and this exposed them to rich encounters with people with whom they typically have very little day-to-day contact. A key element of innovation is the opportunity to come into both random and structured contact with people and events outside of one's everyday frame, and serious games such as 'Broken Cities' are an effective means to do so (see Figure 8.3).

Another robust finding that held up in the games played not only with planners but also with students was that the social nature of the game combined with the large number of possible decisions in any given turn encouraged a complex dialogue between the players (and themselves) about the positive and negative possible consequences of investment decisions. Moreover, players often spent a significant amount of time discussing the fairness of the distributional impacts of one another's investment decisions, as well as having to negotiate policies and regulations to control emissions if they began to rise too quickly.

In the de-briefing sessions that followed the game play, many of the planners expressed satisfaction with the relationship between the necessarily limited realism of the game and the types of conflicts and issues that they encountered in their own work with climate change. 'Broken Cities' has also evolved to encompass critiques from players, including the development of currency and more formal rules governing the regulatory and legal instruments available to the players. Consequently, there has been a double learning between the practitioners and researchers through both the development and the playing of the game.

Analysis

Our results indicate that the use of serious gaming within an action research framework has significant potential to enable co-production of knowledge, as well as to encourage reflexive approaches to learning by doing. As Huntjens *et al.*, (Chapter 2, this volume) note, action research is characterized by its aim to contribute to social action, or in our case to contribute innovative forms of planning within climate change planning. In this regard, action research as a methodology is particularly useful as it is able to capture and reflect on two ideal types of learning and innovation: STI mode (science, technology, and innovation), and DUI mode (doing, using, and interacting) (Jensen *et al.*, 2007). At an organizational level, the possible synergy between these two modes of learning and innovation can perhaps best be described as the necessity to combine knowledge management strategies using ICT as a tool for codifying

Figure 8.3 Example of the game being played

and sharing information with strategies embracing communities of practice and informal communication in order to mobilize tacit knowledge; or, in the action research context, the use of codified global knowledge in the form of a theoretical framework combined with practice and local experience as the primary source of questions and dilemmas.

Limitations of the study

One of the more prosaic problems encountered with action research methods in this case has been access to the planners themselves. For example, day-to-day time constraints, scheduling conflicts, multiple competing strategic and operational aims, shifting political priorities, and conflicts between the normative ideal of the reflective practitioner (Schön, 1983) and the reality of new public management (Sager, 2009) all serve in one way or another to inhibit the ability of the researcher and practitioner to learn from each other, to varying degrees.

One of the methods used to counter these constraints has been to, wherever possible, build reflective space into the existing working practices, but there is a long lead time necessary for this to occur, and it helps immensely to maintain a consistent physical presence within the planning offices themselves.

Another obstacle was to convince the planners that playing games is not a waste of time. Our experience has been that, once players get past the idea that 'it is only a game' and become engrossed in the game itself, there is a clear tendency for them to become much more enthusiastic about the idea of serious games. Only two gaming sessions have been held so far, and it would be useful to have a longitudinal design whereby the same (or similar) groups of planners played the game over a period of years in order to track the ways in which their absorptive capacity and learning potential changed (or stayed the same). The sample size of this study is too small to engender robust generalizations, and it would be preferable to host gaming sessions in a number of different contexts globally in order to understand different social, cultural, geographic, and political drivers of innovation and learning within planning agencies.

One final challenge to mention is the difficulty of managing changes in the organization during the research project. For example, the City of Copenhagen has, from 1 January 2014, merged the mitigation and adaptation planning units into one. From a policy and planning perspective, this is a positive development that allows for a more integrated approach to dealing with climate change and urban planning and is something that we have been advocating in both Portland and Copenhagen for some time. From a research perspective however, it is problematic. This fundamental change in the institutional and political arrangements in Copenhagen requires the action researcher to return to the beginning of the research cycle in order to revise, refine, and reassess the hypothetical bases for the research in the first place. For the planners involved in the research project, this does not present significant problems for them as they are quite accustomed to disjunctive decision and implementation environments, whereas for the researcher the demands of a systematic empirical investigation require a return to step 1. So although action research and case study research allow for the study of contemporary phenomena *in situ*, the mismatches between the logic and needs of the research and practice communities should not be minimized or overlooked. Moreover, it is recommended that researchers develop, as part of their overall research design, contingency plans in order to address some of these challenges before they are encountered in the field.

Conclusions and lessons learned

The initial question that we pursued was the extent to which serious games could stimulate reflection and learning among planners working with climate change mitigation and adaptation. Our study indicates that there is significant promise for action researchers within this field to use serious games to trigger reflection and learning. We found that one of the most powerful drivers of learning is the intense social interaction necessitated by the structure and mechanics of the game, whereby players must think, plan, and act in a dynamic environment of competition and cooperation. The dialogic nature of the game space requires players, who may not necessarily share the same values and priorities, to reach an operational consensus about how to fairly distribute the costs and benefits of both the mitigation and the adaptation goals embedded in the game.

On the basis of our experiences, and those of many others throughout the world, serious games ought to become a standard component of the action research toolkit. The particular combination of simulated real-world dilemmas, intensive social interaction, and the potential for co-production of knowledge make serious games a useful means of developing a more complete understanding of reality. The types of data that are available from serious games, such as role-playing games, card games, board games, or digital games, differ from those acquired using more traditional social science data collection methods such as interviews, surveys, or participant observation because of the real-time dynamics and dialogic nature of the gaming space. In our particular case, serious games created a space for capturing complex social dynamics and verbalized thought processes (talking while doing) that would be difficult to re-create in a normal interview situation.

There are a number of potential avenues for further research, but one area that we shall be developing further is more standard methods for both data collection and analysis through the use of serious games. Measuring learning effects over the short term is difficult, and the extent to which learning translates into changed actions is not readily clear either. Lastly, the extent to which serious games reflect the reality of any given phenomena is still unclear, and also the extent to which this matters to the players themselves. Our research suggests that slavish fealty to the 'real' nature of the problem matters less than capturing the essence of the challenge and, perhaps most of all, having fun.

Notes

1 The synergies, conflicts, and trade-offs elucidated for both Copenhagen and Portland are the results of desk research that analysed successive generations of climate change mitigation, adaptation, energy, natural systems, and comprehensive plans for both cities in the period between 1990 and 2013.
2 'Broken Cities' is freely available for download through a Creative Commons licence and can be found at: http://www.klimalab.dk/#!broken-cities/c20w6

References

Abbott, C. and J. Margheim (2008) Imagining Portland's urban growth boundary: planning regulation as cultural icon, *Journal of the American Planning Association*, 74(2): 196–208.

Abt, C.C. (1970) *Serious games*, New York: Viking Press.

Betsill, M. and H. Bulkeley (2007) Looking back and thinking ahead: a decade of cities and climate change research, *Local Environment*, 12(5): 447–456.

Biesbroek, R., R. Swart and W. van der Knaap (2009) The mitigation–adaptation dichotomy and the role of spatial planning, *Habitat International*, 33(3): 230–237.

Carmin J., N. Nadkarni and C. Rhie (2012) *Progress and challenges in urban climate adaptation planning: results of a global survey*, Cambridge, MA: MIT Press.

City of Copenhagen (2009) *Copenhagen climate plan: the short version*, Copenhagen: City of Copenhagen.

City of Copenhagen (2011) *Copenhagen climate adaptation plan*, Copenhagen: City of Copenhagen.

City of Copenhagen (2012a) *Cloudburst management plan*, Copenhagen: City of Copenhagen.

City of Copenhagen (2012b) *KBH 2025 klimaplanen: en grøn, smart og CO²-neutral by*, Copenhagen: City of Copenhagen.

City of Portland and Multnomah County (2001) *Local action plan on global warming*, Portland, OR: City of Portland.

City of Portland and Multnomah County (2009) *Climate action plan 2009*, Portland, OR: City of Portland and Multnomah County.

City of Portland and Multnomah County (2012) *Climate action plan 2009: year two progress report*, Portland, OR: City of Portland.

City of Portland and Multnomah County (2014) *Climate change preparation strategy: Risks and vulnerabilities assessment*, Portland, OR: City of Portland and Multnomah County.

Cohen, S. and M. Waddell (2009) *Climate change in the 21st century*, Montreal: McGill-Queens University Press.

Greenwood, D.J. and M. Levin (1998) *Introduction to action research: social research for social change*, Thousand Oaks, CA: Sage Publications.

Haug C., D. Huitema and I. Wenzler (2011) Learning through games? Evaluating the learning effect of a policy exercise on European climate policy, *Technological Forecasting and Social Change*, 78(6): 968–981.

Jensen, M.B., B. Johnson, E. Lorenz and B. Å. Lundvall (2007) Forms of knowledge and modes of innovation, *Research Policy*, 36(5): 680–693.

Juhola, S. and L. Westerhoff (2011) Challenges of adaptation to climate change across multiple scales: a case study of network governance in two European countries, *Environmental Science and Policy*, 14(3): 239–247.

Juhola, S., P. Driscoll, J. Mendler de Suarez and P. Suarez (2013) Social strategy games in communicating trade-offs between mitigation and adaptation in cities, *Urban Climate*, 4: 102–116.

Kahan, D.M., E. Peters, M. Wittlin, P. Slovic, L.L. Ouellette, D. Braman and G. Mandel (2012) The polarizing impact of science literacy and numeracy on perceived climate change risks, *Nature Climate Change*, 2(10): 732–735.

Meijer, S. (2010) Gaming simulations for railways: lessons learned from modeling six games for the Dutch infrastructure management, in X. Perpinya (ed.) *Infrastructure design, signaling and security in railway*, 275–294, Rijeka: InTech.

Miljøministeret (2007) *Fingerplan 2007*, København: Miljøministeret.

Næss, P. (2005) Residential location affects travel behaviour – but how and why? The case of Copenhagen metropolitan area, *Progress in Planning*, 63(2): 167–257.

Næss, P. (2011) 'New urbanism' or metropolitan-level centralization, *The Journal of Transport and Land Use*, 4(1): 25–44.

Oresundregion (2013) *Fakta om Øresundregionen*, Malmö: Örestat.

Patt, A.G., N. Peterson, M. Carter, M. Velez, U. Hess and P. Suarez (2009) Making index insurance attractive to farmers, *Mitigation and Adaptation Strategies for Global Change*, 14(8): 737–757.

Peirce, C.S. (1982) *Writings of Charles S. Peirce: a chronological edition*, Bloomington, IN: Indiana University Press.

Reason, P. (1994) *Participation in human inquiry*, London: Sage Publications.

Reason, P. (2003) Pragmatist philosophy and action research: readings and conversation with Richard Rorty, *Action Research*, 1(1): 103–123.

Reckien, D. and K. Eisenack (2013) Climate change gaming on board and screen: a review, *Simulation & Gaming*, 44(2–3): 253–271.

Rijcken, T., J. Stijnen and N. Slootjes (2012) 'SimDelta' – inquiry into an internet-based interactive model for water infrastructure development in The Netherlands, *Water*, 4(2): 295–320.

Rosenzweig, C., W.D. Solecki, S.A. Hammer and S. Mehrotra (eds) (2011) *Climate change and cities: first assessment report of the urban climate change research network*, Cambridge: Cambridge University Press.

Røvik K.A. (1998) *Moderne organisasjoner: trender i organisasjonstenkningen ved tusenårsskiftet*, Oslo: Fakbokforlaget.

Sager, T. (2009) Planners' role: torn between dialogical ideals and neo-liberal realities, *European Planning Studies*, 17(1): 65–84.

Schön, D. (1983) *The reflective practitioner: how professionals think in action*, New York: Basic Books.

Statistics Denmark (2013) *Statistisk årbog 2013*, Copenhagen: Statistics Denmark.

Swart, R. and F. Raes (2007) Making integration of adaptation and mitigation work: mainstreaming into sustainable development policies? *Climate Policy*, 7(4): 288–303.

Tol, R.S.J. (2005) Adaptation and mitigation: trade-offs in substance and methods, *Environmental Science & Policy*, 8(6): 572–578.

Ulrich, M. (1997) Games/simulations about environmental issues: existing tools and underlying concepts, in J. Geurts, C. Joldersma and E. Roelofs (eds) *Gaming/simulation for policy development and organizational change: proceedings of the 28th Annual Conference of the International Simulation and Gaming Association*, 301–311, Tilburg: Tilburg University Press.

United States Census Bureau (2010) Guide to state and local census geography, retrieved from http://www.census.gov/geo/reference/geoguide.html (accessed 20 April 2010).

Wilbanks, T.J. and J. Sathaye (2007) Integrating mitigation and adaptation as responses to climate change: a synthesis, *Mitigation and Adaptation Strategies for Global Change*, 12(5): 957–962.

9 Using role-play simulations to encourage adaptation

Serious games as tools for action research

Todd Schenk and Lawrence Susskind

Introduction

The first chapter of this book made the case that action research has a lot to offer those trying to encourage adaptation to climate change. The technical difficulties of formulating and interpreting climate change forecasts are substantial, but not nearly as challenging as the governance problems that arise when collective risk management choices must be made (Lemos *et al.*, 2012; Moser and Ekstrom, 2010). Communities must take action, but no single actor alone can address the risks posed by climate change. Effective adaptation requires scientists, engineers, designers, policymakers, and stakeholders to work together (Birkmann *et al.*, 2010). Most groups do not know how to do this. Action researchers can be of assistance. They can clarify differing perceptions of risk and competing estimates of costs and benefits. They can also help decision makers and other stakeholders strengthen their collaborative decision-making capabilities.

Action research calls on scholars to engage in the 'practice of participation, engaging those who might otherwise be subjects of research or recipients of interventions to a greater or lesser extent as inquiring co-researchers' (Reason and Bradbury, 2008: 1). In an action research context, researchers and their partners collectively analyse a situation and interpret findings together. The intent of their joint inquiry is to produce widely supported proposals for action – and ultimately concrete outcomes – that address collectively identified needs or problems. This is exactly the kind of help that officials need.

We make the case in this chapter that role-play simulation (RPS) exercises – a type of serious game – can be a tool for helping public officials educate both others and themselves about climate change risks (Susskind and Schenk, forthcoming). They can also be used to help elected and appointed officials get a sense of what citizens and other stakeholders are worried about and why. RPSs can be used to bring researchers, policymakers, and a range of stakeholders together to collectively wrestle with fictional yet realistic decision-making challenges in a low-cost, low-risk setting (Schenk, 2014; Susskind and Rumore, 2013). When the playing of serious games precedes actual decision making, stakeholders and decision makers can use the experience to prepare for the governance challenges they face. The cases presented in this chapter illustrate how stakeholders and researchers have used RPSs to anticipate the conflicts likely to arise when collective climate risk management decisions must be made, and to explore how they might be overcome.

This chapter draws on our experiences of using RPSs around the world to help communities and institutions advance climate adaptation and risk management efforts while simultaneously advancing the theory of collaborative adaptive management. We spell out the benefits and limitations we have found of using RPSs (or serious games) as a tool for action research. We base our conclusions on what we have learned from using RPSs in climate adaptation planning in coastal New England (United States) and with infrastructure planners and decision makers in Singapore and Rotterdam.

Serious games and role-play simulation exercises

Serious games engage players in trying to solve fictional challenges within clear constraints (Abt, 2002). Although serious, they are enjoyable ways of learning complex material and new ways of making decisions (Abt, 2002). Serious games have been used in a wide range of situations, ranging from elementary school classrooms to military strategic command centres (Abt, 2002). Serious games have proved useful in a wide range of public policymaking contexts (Dolin and Susskind, 1992; Mayer, 2009; Najam, 2001). Recently, they have begun to gain traction as tools for building the capacity of decision makers and stakeholders facing difficult climate change adaptation choices (Mendler de Suarez *et al.*, 2012; Plumb and Schenk, 2011; Schenk, 2014; Susskind and Rumore, 2013; Susskind and Schenk, forthcoming).

Serious games range from simple exercises that can be introduced and played in just a few minutes with a few props and no written instructions, to technically complex exercises that stretch over days and involve substantial background reading and sometimes online computational support. The 'Humans vs. Mosquitoes' game is an example of the former. It is a fast-paced game that requires only a few tokens and can be played in ten minutes (Mendler de Suarez *et al.*, 2012). Nonetheless, it can provide vulnerable communities with a clear sense of how the risks of dengue fever might shift because of climate change. At the other extreme, the 'Indopotamia' RPS is usually played over four two-hour segments and requires extensive preparatory reading (Islam *et al.*, 2012). It is used to show water professionals from around the world how climate change might need to be taken into account in resolving a wide range of transboundary water disputes.

RPSs are a type of serious game in which multiple participants engage in mock decision making bounded by confidential role instructions. As Figure 9.1 illustrates, RPSs resemble traditional games in that they have fixed rules, constrained outcomes, and offer interactive ways of mastering a great deal of technical information. The basic rules and contextual constraints are presented in *general instructions*, which all players receive. Participants are assigned roles and given *confidential instructions* to help them play unfamiliar parts in a realistic

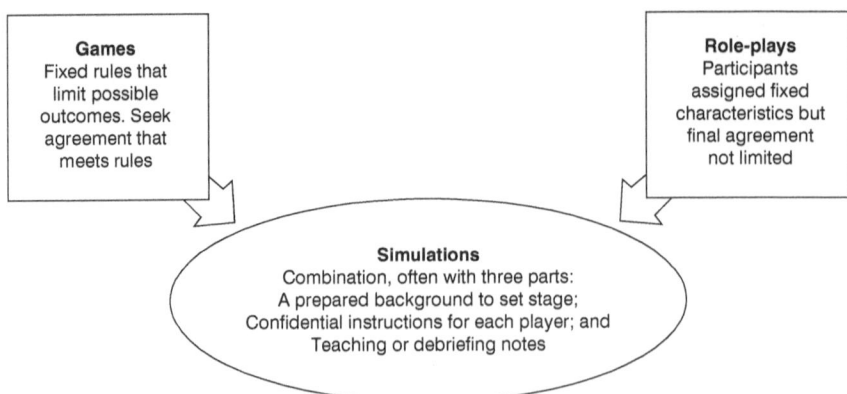

Figure 9.1 Characteristics of role-play simulation exercises (source: adapted from Susskind and Corburn, 1999)

fashion. The confidential instructions for each role are often based on what actual individuals in those real-life roles have had to say about situations similar to those portrayed in the game.

Debriefings are central to RPS exercises. Facilitators help participants reflect on their game experience and draw connections between what happened and the situations they face in their own lives. Debriefings are especially instructive when several groups play the same game at the same time so there are multiple results to compare. Before-and-after surveys, the recording and analysis of exercises and debriefings, and follow-up interviews can turn these exercises into research tools. If game players spell out their understandings and expectations before and after participating in an RPS, a government trying to work with stakeholders to advance an adaptation planning effort can find out what the key obstacles are likely to be and how they might be overcome. When this kind of investigation follows action research prescriptions, the findings can enhance the political legitimacy of the policies and programmes that governments decide to pursue.

Role-play simulation exercises as part of action research

Action research involves cycles of action and reflection (Reason and Bradbury, 2008). Carefully tailored games can help participants understand the underlying forces that will have to be dealt with in practice before any change is possible. They can also share the scientific or technical materials that stakeholders will need to master before they can help to generate technically credible proposals. As Figure 9.2 illustrates, RPSs aim to provide opportunities for communities to reflect on the problems they wish to address, and what the obstacles might be to taking action.

RPSs are particularly appropriate at the front-end of adaptation planning efforts, when stakeholders are just beginning to explore the climate risks they face, as well as the obstacles likely to inhibit collective risk management efforts

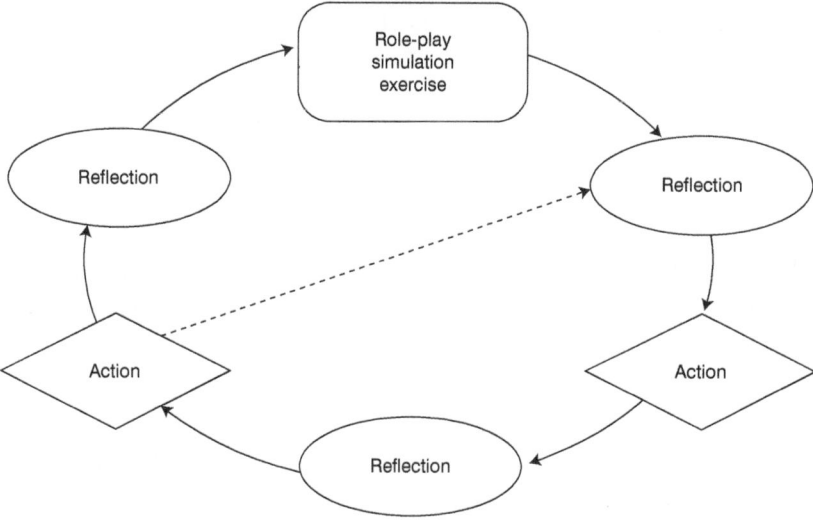

Figure 9.2 Role-play simulation exercises in the action–reflection cycle

(Susskind and Schenk, forthcoming). RPS exercises prime stakeholders for action by immersing them in situations similar to their own, but sufficiently simplified so that they can see what the underlying governance dynamics are that must be dealt with. This also makes RPSs appropriate for theory building around governance, because researchers can learn with multiple stakeholders in multiple communities as they piece together ways of initiating adaptation in different kinds of places.

Using the action research taxonomy introduced in Chapter 2, RPS exercises can play integral roles in encouraging *action science* and *cooperative enquiry* (Reason and Bradbury, 2008). They can assist in cooperative enquiry when they involve stakeholders in the scoping, design, and implementation of RPSs. The same exercises can contribute to action science when they challenge stakeholders to confront the gap between their *theories-in-use* and their *espoused theories*, thus encouraging them to engage in systematic reflection.

From a research perspective, RPS exercises have contributed to theory building in the social sciences (Najam, 2001). RPS exercises allow public policy scholars and other social scientists to closely observe what happens and what is said during and after play. Scholars monitor how events unfold in exercises, record participants' reflections during the debrief conversations, and collect further information via surveys and follow-up interviews. For example, we have used the 'Mercury Game' to help scientists confront the difficult choices facing diplomats trying to reach agreement on the terms of a global treaty restricting mercury in the environment.[1] Game results can be used to build theory around the ways in which science-intensive policy questions are handled. All of this can be done at relatively low cost and in a politically and psychologically safe setting. Exercises also allow for rapid experimentation, yielding insights more quickly than could otherwise be gathered from analogous real-world situations. Rather than confronting their own adaptation challenges at the outset, stakeholders, decision makers, and researchers can ease into discussions of potential obstacles and difficulties working from hypothetical but realistic cases. Of course, it is imperative that skilled facilitators be able to challenge participants to draw the connections between RPS results and their own real-world situations.

Given the aforementioned characteristics, RPS exercises are particularly useful in the context of climate change adaptation. They allow individuals and groups to wrestle with complex climate-related challenges in a simplified form. They also allow people to experiment with tools and approaches to managing risks collectively, without anyone having to make actual political commitments before they are ready, and they provide an opportunity for stakeholders to experiment with new forms of decision making that would be 'hard to sell' in real life.

The remainder of this chapter explores the benefits and challenges of using RPS exercises in the context of both promoting and understanding climate change adaptation. We illustrate the use of RPSs as an action research tool by examining two cases.

New England Climate Adaptation Project

The New England Climate Adaptation Project (NECAP) is an action research project involving the MIT Science Impact Collaborative, the Consensus Building Institute, the National Estuarine Research Reserve System (NERRS), and four communities on the northeast coast of the United States. The project aims to increase the capacity of cities and towns in the region to identify the risks posed by climate change and to build public support for collaborative efforts to manage them (NECAP, 2013; Rumore, 2014; Susskind and Rumore, 2013). These goals are clearly both action and research oriented.

The NECAP project began with the assumption that coastal communities must do something to manage climate risks. To do this effectively, they must involve all relevant stakeholders, because any assessment is as much about risk perception as it is about quantitative estimates of the damage climatic changes might cause. With risk assessments in hand, there is still a problem of deciding how to manage those risks (i.e. what actions to take). This is a collective decision-making problem. For community risk management choices to be credible and widely supported, it is important that stakeholders believe they have played a genuine part in decision making (Edelenbos *et al.*, 2010; Fung, 2004). Action research processes can be used to build public awareness of climate risks and engage communities in considering how they might manage them.

RPS exercises can be used to educate communities about the climate-related problems and opportunities they face. When the debriefings are taken seriously, exercises can help officials to understand the concerns and possible reactions community members might have to adaptation options identified. Furthermore, the process of participating can increase the overall level of optimism and support within communities for taking action. RPS exercises are at the core of the NECAP. Individual exercises were developed for each of the four partner municipalities. They were run with hundreds of stakeholders, including business interests, coastal residents, and environmentalists (NECAP, 2013). Public officials played integral roles throughout, including helping to design the simulations and putting together lists of actions that might be taken.

The exercise developed for use in Barnstable, Massachusetts, is called 'Coastal Flooding in Shoreham: Responding to Climate Change Risks' (Agatstein, 2013). The fictional town of Shoreham has a problem: the frequency and intensity of storm surges appear to be increasing, damaging both public and private property. Some stakeholders assert that climate change is to blame, and many demand that the municipality do something about it. The town manager has subsequently created a Coastal Flooding Task Force to investigate how climate changes might affect Shoreham, and what the community might do to enhance its resilience. The task force includes the assistant town manager, a town planner, an environmentalist, a representative of the business community, the president of a civic association, and a realtor (real estate agent) (Agatstein, 2013). They each have their own perspectives on the nature of the problem and what should be done about it, which, unsurprisingly, reflect the various interests of the constituencies they represent. The task force has been asked to suggest how the town should handle increasing coastal flooding risk, either by building new flood protection infrastructure, flood-proofing existing structures, or imposing new land-use restrictions. The task force can recommend tackling flooding by modifying its zoning ordinances to reduce future development in the 100-year or 500-year flood zones, subsidizing the flood-proofing of housing in the 100-year flood zone, or allowing at-risk properties to armour their section of the coast (Agatstein, 2013). Each option comes with its own set of costs and benefits, and is more or less appealing to particular actors. The task force's goal is to reach an agreement that all members can support.

The Shoreham RPS does not present an exhaustive list of possible options that the actual town of Barnstable might use; nor does it enumerate all the costs and benefits associated with each course of action. The value of the RPS is in demonstrating that there are multiple options that come with differing costs and benefits, and that different segments of the town will prefer different policies and actions. Participants experience first-hand that adaptation is not just a technical problem, but a governance challenge as well. The RPS is meant to help those playing the game appreciate the governance challenges involved, and to subsequently feel more empowered to address them.

From a research perspective, the exercises are followed by detailed debrief conversations in which participants can reflect on what happened and why, and whether or not (and how) the outcomes relate to the research questions posed. Participants also complete pre- and post-exercise surveys and a subset are interviewed to assess whether and how the exercise shifts their perceptions of, and response to, potential climate threats. Finally, polling throughout the community is conducted to assess if and how the exercises shift wider perceptions throughout the community.

Fictional Shoreham has a lot in common with real-world Barnstable, while not being quite the same. The RPS is based on community-scale climate projections prepared with great care by the scientists and engineers on the project team, along with extensive interaction with municipal officials (NECAP, 2013). The simplification of the Barnstable story is meant to give participants a bit of distance from the politics and particulars of their actual town. It allows game players to focus on the most important issues while pushing other things into the background. It presents both the technical risk assessment and the conflicting interests among various stakeholder groups.

The NECAP project is not simply presenting climate data in an innovative way. 'The idea was to take a look at some specific attitudes about climate change … and start a conversation about climate change based on real information', said project partner Elizabeth Jenkins from Barnstable's Growth Management Department (MacDonald, 2014). Although the climate projections are valuable, the 'social data' are just as important. Project partners in Barnstable sought to 'take the temperature' of the community before beginning to talk about actual expenditures or regulatory changes. They partnered with the NECAP research team because they thought the exercise could help them achieve these goals (MacDonald, 2014). From an action research perspective, the exercise and the before-and-after surveys of game participants along with a random sample of residents gave the town a strong scientific basis on which to proceed, and built public awareness of a problem requiring community-wide action.

RPS participants are intentionally assigned roles different from those they hold in the real world. This is done so that they can gain a deeper appreciation of the perspectives of others and begin to think about how they might meet others' interests while also advancing their own. Ed Dewitt, the executive director of the environmentally oriented Association for the Preservation of Cape Cod and a participant in the Shoreham exercise, reflected positively on this feature of the Barnstable game, stating that 'It's always a healthy exercise when we step outside of our normal roles … I think there [are] probably times when it's good for me to see the world through the eyes of a taxpayers' association' (MacDonald, 2014). From an action research perspective, fostering such reflectiveness can be very powerful.

The similarities between the RPS exercises and the real communities on which they are based are not lost on those participating. Aside from the obvious parallels between the fictitious and real-world communities, debriefings immediately follow each simulation, during which participants are encouraged to reflect on the results of the game, the differences in the results that other groups produced playing the same game in the same time period, and how the results relate to their own situations. In Barnstable, 45 per cent of game players had little or no confidence before the game began that their town might be able to take action to reduce the community's vulnerability to climate risks. After just one hour of role-play and a half hour debriefing, 23 per cent felt more confident.[2]

The NECAP team met with elected officials and other community leaders multiple times before and after the game was played to see whether the town was ready to undertake in real life a collaborative adaptation planning effort like the one the participants in the fictitious Shoreham had completed (NECAP, 2013; Rumore, 2014; Susskind and Rumore, 2013). The team is also monitoring the meetings and activities of agencies on an ongoing basis to see whether or not there has been any tangible action on climate change adaptation prompted by publicity about the exercises. Although only 150 residents participated directly in the RPS sessions, the expressed interest in follow-up activities in the partner municipalities suggests that they did foster action. At time of writing, the NECAP team is waiting to see whether an independent poll of a random sample of residents in Barnstable provides evidence of a wider shift in public opinion.

Preliminary findings suggest that participation in RPS exercises can significantly increase stakeholder awareness of climate change risks and generate optimism about taking collective action in response to them (Susskind and Rumore, 2013). In Barnstable, 22 per cent of participants said the exercise affected their views about climate change, and 57 per cent said it is 'very important' that 'residents, local groups, and businesses be involved in deciding how to respond to climate change risk'. There is also evidence from the NECAP project that most people who played the game are more optimistic afterwards that their local government could, in fact, facilitate agreement among stakeholders on what might be done to manage climate-related risks.[3]

Action researchers begin with the goal of helping actual communities chart the social changes they want to try to achieve. In the NECAP case, the products of the RPSs are increased optimism among residents that informed collective agreements can be reached in the face of enormous uncertainty about how to deal with climate change, enhanced understanding of the risks and potential adaptation options, and a sense of how adaptation efforts might be phased to prepare for the future while acknowledging uncertainty.

Infrastructure planners and decision makers in Singapore and Rotterdam

Climate change poses a range of threats to transportation infrastructure. The data and tools for analysing these threats are continuously improving, but significant governance challenges remain. It is unclear how the complex institutions that control infrastructure planning can be helped to reach informed agreements. Dutch organization TNO is supporting the 'Harbouring Uncertainty' project as part of the wider Knowledge for Climate programme because it wants to work with government agencies in the Netherlands and beyond to help them expand their understanding of how to win support for long-term planning efforts even in the face of substantial uncertainty.

The project involves stakeholders in Singapore and Rotterdam exploring the use of multiple scenarios and multi-stakeholder dialogue as means of injecting greater concern about risk and uncertainty into infrastructure planning. Key actors in relevant agencies in both metropolitan areas are directly involved. Project participants are engaged via an RPS exercise called 'A New Connection in Westerberg'. Westerberg is a large port city that is planning to build a new motorway to alleviate major traffic congestion (Schenk, 2013). Unfortunately, a new climate impact assessment just released suggests that the road – along with the existing motorway it is meant to complement and other infrastructure throughout the city – could be vulnerable if the wrong design and route decisions are made. Changing the design and/or the route may alleviate some climate risks, but will generate other costs and trade-offs. The Transportation Agency has convened a carefully selected group of stakeholders from both within and outside government to consider the risks and trade-offs. The exercise assigns the participants to specific roles at a third meeting of the evaluation group. They are asked to generate a consensus recommendation.

The options before the group include: building a below-grade road; elevating the road to protect it from flooding; diverting the road away from the city through a wetland; or shelving the new road and investing funds in expanding the capacity of, and climate-proofing, the existing road (Schenk, 2013). Freight and passenger rail options also exist, but are only mentioned to some of the role players via their confidential instructions. Different options are more or less appealing to different actors, depending on their perspectives, priorities, and interests. For example, the environmentalist is adamantly opposed to the alternate route through the wetland because of the negative impacts it would probably have on important ecosystems. She or he is concerned about climate change, but feels that the best option would be to strengthen existing road and rail infrastructure rather than build a new motorway. Conversely, the port representative is not as concerned about climate change, but supports the modified wetland route because it offers a direct connection between the harbour and the rest of the country's road network.

The exercise helps participants explore, in a relatively safe environment, options they might consider in the real world. There will inevitably be more options and a wider set of trade-offs in the real world, but the process of exploring and evaluating alternatives is initiated via the exercise. As in the NECAP project, simplification allows the group to focus on decision-making dynamics, but diminishes opportunities to delve into substantive adaptation options in more detail. Abstraction creates a safe space for discussion, but makes the task of relating lessons learned back to practice more difficult.

There are two versions of 'A New Connection in Westerberg'. The first offers a traditional risk assessment forecast. The second provides four qualitative scenarios (i.e. divergent possible futures). Figure 9.3 illustrates the four scenarios. These two versions are being run with parallel groups in each city to see if, and how, the availability of contingent scenarios produces results different from those produced by groups given a definitive forecast. Planners and decision makers often express support for the idea of using scenarios in situations characterized by a high degree of uncertainty, as is the case with climate change, but it is difficult in practice to incorporate multiple scenarios into decision making (Lempert, 2013; Zegras and Rayle, 2012) .

Our experiences with stakeholders in Rotterdam and Singapore suggest that multiple scenarios sometimes produce paralysis, leading participants to conclude

Wet and quiet	Wet and busy
• Precipitation and/riverine flooding leads to higher water level in the near future • Vehicular traffic volume remains constant or declines in the coming years	• Precipitation and/riverine flooding leads to higher water level in the near future • Vehicular traffic increases steadily and substantially in the coming years
Dry and quiet	Dry and busy
• Slow or no increase in precipitation and flooding risks • Vehicular traffic volume remains constant or declines in the coming years	• Slow or no increase in precipitation and flooding risks • Vehicular traffic increases steadily and substantially in the coming years

Figure 9.3 Scenarios (i.e. possible futures) from 'A New Connection in Westerberg' role-play simulation exercise (source: Schenk, 2013)

that the experts need to work harder to devise a better (single) estimate of what is likely to happen. Stakeholders that might benefit if decision making is delayed often use uncertainty as a stalling tactic. In both Singapore and Rotterdam, participants looking to make decisive decisions reverted to using single forecasts by focusing on only one scenario. Some participants tried to protect themselves by using the worst-case scenario. Scenario planning may be a good idea, but more attention must be paid to helping decision makers cope with uncertainty and multiple futures. In addition to informing our research, participation in the RPS appears to be encouraging participants to think about how they might use scenario planning as their agencies ramp up their adaptation efforts.

Enhancing the rail network is a viable way to alleviate traffic congestion while also climate-proofing even more of Westerberg's infrastructure (Schenk, 2013). However, it is not presented as an option in the general instructions. Instead, the environmental activist at the table is guided to promote it, calling on the Transportation Agency's technical expert for information. The expert has data on expected costs and benefits in his or her private instructions, but is guided only to share them when asked. This feature was built into the game to see what happens when ideas come from a single unconventional and potentially marginal source. Interestingly, the idea got a sound hearing in virtually every group, and was chosen many times. This suggests that stakeholders from outside government can play instrumental roles in collaborative planning processes, contributing ideas and perspectives that help shape the outcomes.

Reflecting on the role the environmentalist played in advocating an alternative approach to decision making, participants in both Singapore and Rotterdam noted the value of involving multiple stakeholders at all stages of infrastructure planning and decision making. Participants also realized that accomplishing this is not easy for many reasons: first, a great deal of the groundwork in planning and decision making is informal, making it difficult to involve actors from outside the traditional network of civil servants who know and trust one another. Second, the belief in rational decision making persists; many civil servants believe that only technical experts can and should find the *right* solution, whereas other stakeholders simply add bias to the proceedings, resulting in poorer outcomes. In Singapore, there is fear among civil servants that expanding conversations around things like climate change adaptation will reveal that experts do not have all the answers, eroding faith in government. Finally, it is not always clear how stakeholders can and should be involved, especially from a legal perspective. Although there were some exceptions, most participants in the RPS exercise concluded that involving stakeholders can lead to wiser decisions, and agreed that climate adaptation is as much a governance as a technical problem, but there is neither experience nor clear thinking about how practice should change. This represents an area of governance in need of more attention.

Participants suggested that they learned a great deal by trying to think like someone else; they gained a new appreciation for other stakeholders' perspectives and interests, and are now more inclined to look for creative options that serve multiple interests concurrently. RPS participants with more technical positions in their real-world agencies note that they have gained a new appreciation for the governance challenges and competing interests that must be balanced. Conversely, players from policy and political backgrounds gained new appreciation for what technical uncertainty means and why they may want more than best-estimate forecasts when considering highly uncertain factors like future climate conditions. This newfound appreciation for the interests and perspectives of other stakeholders was put to good use immediately, as stakeholders moved from participating in the RPS to brainstorming ways of addressing adaptation options in their work.

Data are collected in various ways. The runs of 'A New Connection in Westerberg' are video-recorded for later coding and analysis, following established qualitative research practices. Pre- and post-exercise surveys, focus-group-like debriefings, and in-depth one-on-one interviews are also being used to improve our understanding of what happens and why. The 'Harbouring Uncertainty' project seeks to generate insights that can inform practice around the governance of climate adaptation beyond the cities being studied, while also having a meaningful influence on those directly involved.

Discussion: the benefits and challenges of using RPS exercises in the context of action research aimed at encouraging adaptation planning

As the above cases illustrate, RPSs are useful in action research efforts. Their practical value is largely educational, but careful reflection on the results of serious games can stimulate collective action that might not otherwise happen. Researchers can also extract wider insights from what happens during exercises, and what is said before and after play.

Through RPS exercises, participants master a great deal of technical information in a relatively painless fashion and a short amount of time. For example, the NECAP exercises give participants simplified climate projections very similar to those compiled for their own municipalities. The data become much more tangible and understandable vis-à-vis the discussions that evolve within the exercises. Participants gain fresh perspectives and a greater appreciation of the views of others. RPS exercises provide opportunities for participants to grapple with complex issues like climate change, experimenting with tools and considering possible action strategies without having to make actual commitments. RPSs immerse participants in situations, engaging them emotionally by asking them to act out realistic situations; this can be much more powerful than traditional passive education and research techniques (Susskind and Rumore, 2013).

Our experiences suggest that this kind of learning may be particularly powerful when the goal is to translate new knowledge into action in collective decision-making contexts. Exercise participants do not just increase their own understanding. They increase their appreciation for the multifaceted nature of decision making in complex environments, and thus their ability to work with other stakeholders to enhance resilience and address vulnerabilities. They are particularly useful for facilitating agenda setting early in governance processes around complex matters like climate adaptation. Exercises may be used to initiate new adaptation efforts, as is the case in the NECAP project. At the same time, researchers can use the cumulative results of multiple runs of the same role-play to learn how a particular community is leaning with regard to a particular policy question.

RPS exercises are most effective when they are driven by explicit research goals or learning objectives. For example, the NECAP exercises were devised with the learning goal of introducing specific climate risks, adaptation options, and potential costs and benefits associated with these options to stakeholders in each community. From a research perspective, the goal is enhancing understanding of the efficacy of RPSs for facilitating action on adaptation. Community partners played a central role in identifying which vulnerabilities their respective exercises should focus on, and how the issues should be framed (Rumore, 2014). Ideally, stakeholders are involved at all stages, from scoping the focus of an RPS to interpretation of the results. In general, such an action research approach means that community partners are involved in a truly cooperative enquiry. Of course, not all participants will want to be intimately involved at every stage. A handful of project partners should be involved throughout, with others engaged on a more limited basis.

It is important to have the relevant stakeholders at the table when the RPS is actually played to maximize the direct benefits exercises bring to communities. This also enhances their use as a research tool. Stakeholders need to see how exercises can help to advance their climate adaptation efforts. The research-oriented 'Harbouring Uncertainty' project has been successful in large part because host organizations in both countries were able to use the games to bring key stakeholders together to explore their shared interests in climate change adaptation.

The design of RPS exercises is critical to their success. The cases discussed here and our wider experiences suggest that basing exercises on analogous real-world situations while abstracting to provide a bit of distance works best. Putting stakeholders into a fictionalized world gives them freedom to explore issues and options without the usual political constraints. However, there is an inherent tension. Abstraction creates beneficial distance from the real world, but makes it harder to translate any lessons learned back to participants' own situations. We spend considerable time and resources ensuring that our exercises are grounded in reality and will resonate with participants. The NECAP exercises are based on, among other things, climate forecasts prepared for each of the partner towns by respected climate scientists (Susskind and Rumore, 2013).

Exercises are necessarily simplifications of reality. Because of limited time and cognitive capacity, participants are presented with limited sets of options and supporting information compared to what they would have to work with in analogous real-world situations. The goal is not to mirror exhaustive processes just like they would play out in the real world, but to provide vivid snapshots of some of the key issues and dynamics that might arise. Here too, there is a trade-off. The simplification process can limit creativity. The debriefing conversations provide opportunities for participants to discuss options beyond those incorporated into the exercise and bring some of the other obstacles they may face back into the conversation. The simplification involved is one reason why exercises may be more appropriate during the earlier agenda-setting stages of adaptation efforts.

A key feature of RPS exercises is that participants are assigned roles other than those they fill in the real world. From a learning perspective, this is central, as participants gain new appreciation for the perspectives and interests of other stakeholders. Participants consistently reflect that this is very enlightening and changes how they approach negotiations afterwards. From a research perspective, the benefits are less clear. In fact, this shift of roles within the games means that actors are not being themselves, reducing what can be learned from watching them. This makes the debrief conversations and follow-up interviews all the more important, as that is when researchers can invoke reflection on what happened in the exercise compared to what happens (or might be expected) in the real world.

RPSs are certainly not always the product of, or contributors to, action research. Exercises are regularly devised for, and used with, stakeholders without their involvement in scoping, design, and interpretation of outcomes. On the other hand, exercises are often used as part of consulting practice to help communities explore their challenges and take action without any wider research implications. Our cases have shown that RPSs are particularly effective when developed in partnership with stakeholders, and when stakeholders play a central part in the interpretation of results. At the same time, wider research findings can be drawn from our experiences working with communities using exercises. Researchers can learn from what happens both within and around the exercises, contributing to the academic discourse in important areas like climate governance. For example, the NECAP project is enhancing our understanding of the role of shared perceptions of what is possible throughout communities before they can most effectively act in situations with significant uncertainty. On the other hand, exercises may not be as useful in well-established policy domains, where more nuanced and technical examination is required.

Social scientists engaged in action research often struggle to produce knowledge that is legitimate in the eyes of their research peers while they are trying to help a particular community wrestle with difficult problems, like how to handle climate change risks. RPS exercises accelerate the cycles of action and reflection by providing stakeholders and researchers with space for experimentation. Groups test options and assess approaches, quickly iterating and revising as they go, without going through the arduous process of implementing real policy experiments. For example, the 'Westerberg' exercise facilitates experimentation with the use of multiple scenarios. However, stakeholders and researchers must be honest about the challenges they are likely to face when translating exercise outcomes to the real world.

Researchers can ensure that they are meeting rigorous standards by accompanying RPS exercises with well-designed surveys, debriefing sessions, interviews, and other qualitative and quantitative research tools. Exercises can also be video-recorded, as is being done in the 'Harbouring Uncertainty' project, for later coding and analysis.

Conclusion

Adaptation is at least as much a governance challenge as a technical one. It requires new ways of incorporating scientific information into political decisions in the face of persistent uncertainty. It requires collaboration among decision makers, technical and scientific experts from different fields, and other stakeholders. Researchers interested in supporting effective adaptation can work with communities to help them explore their vulnerabilities, and how these might be addressed. Action researchers can ensure that community members are not just the subjects of this research, but share in its design and the interpretation of results.

This chapter explored the benefits and limitations of RPS exercises as a tool for supporting action research. Exercises provide opportunities for rapid experimentation and learning. In the spirit of cooperative inquiry, researchers work *with* partners to develop exercises that help them collectively explore emerging risks, opportunities, tools, and perspectives. Exercises are used to explore alternative approaches to decision making and decision-support tools like scenario planning. In the spirit of action science, researchers also gain new appreciation for the multiple perspectives and interests that different stakeholders bring to the table. Participants are challenged to critically reflect on their theories and actions as they devise new approaches at the table. Researchers and community partners learn from what happens during the exercises, and by surveying, interviewing, and debriefing with participants. RPSs do not need to be embedded within action research. However, an action research orientation that involves stakeholders at all stages, from design to interpretation, leads to better RPSs and more effective use of them.

The insights gleaned from RPS exercises can be translated into improved processes and wiser decisions in the real world, and to improved understanding of what constitutes effective adaptive governance in the face of complex risks like climate change. However, the limitations outlined in this chapter must also be recognized and appreciated.

Notes

1 The 'Mercury Game' and many other RPSs are available via the Program on Negotiation at Harvard Law School at www.pon.harvard.edu/store.
2 For more information on lessons learned, please see the NECAP website: http://necap.mit.edu.
3 Thirty-five per cent of participants from the four partner towns felt more confident in their respective town's ability to prepare for climate change after play, whereas only 13 per cent felt less confident. The remaining 52 per cent felt neither more nor less confident.

References

Abt, C.C. (2002) *Serious games*, Lanham, MD: University Press of America.

Agatstein, J. (2013) *Coastal flooding in Shoreham: responding to climate change risks*, Cambridge, MA: Consensus Building Institute, Massachusetts Institute of Technology, New England Climate Adaptation Project.

Birkmann, J., M. Garschagen, F. Kraas and N. Quang (2010) Adaptive urban governance: new challenges for the second generation of urban adaptation strategies to climate change, *Sustainability Science*, 5(2): 185–206.

Dolin, E.J. and L.E. Susskind (1992) A role for simulations in public policy disputes: the case of national energy policy, *Simulation & Gaming*, 23(1): 20–44.

Edelenbos, J., N. van Schie and L. Gerrits (2010) Organizing interfaces between government institutions and interactive governance, *Policy Sciences*, 43(1): 73–94.

Fung, A. (2004) *Empowered participation: reinventing urban democracy*, Princeton, NJ: Princeton University Press.

Islam, S., L.E. Susskind and C.M. Ashcraft (2012) The Indopotamia role-play simulation, in S. Islam and L.E. Susskind, *Water diplomacy: a negotiated approach to managing complex water networks*, 269–314, Washington, DC: Resources for the Future Press.

Lemos, M.C., C.J. Kirchhoff and V. Ramprasad (2012) Narrowing the climate information usability gap, *Nature Climate Change*, 2(11): 789–794.

Lempert, R.J. (2013) *Climate scenarios that illuminate vulnerabilities and robust responses*, *Climatic Change*, 117(4): 627–646.

MacDonald, B. (2014) Climate-change games, *ecoRI News*, 4 January, retrieved from: http://www.ecori.org/massachusetts-climate/2014/1/4/climate-change-games.html (accessed 9 January 2014).

Mayer, I.S. (2009) The gaming of policy and the politics of gaming: a review, *Simulation & Gaming*, 40(6): 825–862.

Mendler de Suarez, J., P. Suarez, C. Bachofen, N. Fortugno, J. Goentzel, P. Gonçalves, N. Grist, C. Macklin, K. Pfeifer, S. Schweizer, M. van Aalst and H. Virji (2012) *Games for a new climate: experiencing the complexity of future risks*, Pardee Center Task Force Report, Boston, MA: The Frederick S. Pardee Center for the Study of the Longer-Range Future.

Moser, S.C. and J.A. Ekstrom (2010) A framework to diagnose barriers to climate change adaptation, *PNAS*, 107(51): 22026–22031.

Najam, A. (2001) Getting beyond the lowest common denominator: developing countries in global environmental negotiations, PhD Dissertation, Massachusetts Institute of Technology.

NECAP (2013) *New England Climate Adaptation Project*, retrieved from: http://necap.mit. edu (accessed 8 January 2014).

Plumb, D. and T. Schenk (2011) Best practice lessons: role-play simulations and managing climate change risks, retrieved from: http://www.cbuilding.org/tools/bpcs/roleplay-simulations-and-managing-climate-change-risks (accessed 7 January 2014).

Reason, P. and H. Bradbury (2008) *The Sage handbook of action research: participative inquiry in practice*, 2nd edn, London: Sage.

Rumore, D. (2014) Building the capacity of coastal communities to adapt to climate change through participatory action research: Lessons learned from the New England Climate Adaptation Project, *Carolina Planning Journal*, Summer 2014.

Schenk, T. (2013) *A new connection in Westerberg: role-play simulation exercise*, Cambridge, MA: Massachusetts Institute of Technology.

Schenk, T. (2014) Boats and bridges in the sandbox: using role-play simulation exercises to help infrastructure planners prepare for the risks and uncertainties associated with climate change, in A.V. Gheorghe, M. Masera and P.F. Katina (eds) *Infranomics: sustainability, engineering design and governance*, 239–255, Berlin: Springer.

Susskind, L.E. and J. Corburn (1999) *Using simulations to teach negotiation: pedagogical theory and practice*, PON Working Paper 99-3, Cambridge, MA: Program on Negotiation at Harvard Law School.

Susskind, L. and D. Rumore (2013) Collective climate adaptation: can games make a difference? *Solutions Journal*, 4(1). http://thesolutionsjournal.com/node/2021

Susskind, L. and T. Schenk (forthcoming) Can games really change the course of history? *Négociations*.

Zegras, C. and L. Rayle (2012) Testing the rhetoric: an approach to assess scenario planning's role as a catalyst for urban policy integration, *Futures*, 44(4): 303–318.

10 Adaptive flood risk management for unembanked areas in Rotterdam

Co-creating governance arrangements for local adaptation strategies

Arwin van Buuren, Mike Duijn, Ellen Tromp and Peter van Veelen

Introduction

Changing climate conditions have major consequences for the unembanked parts of cities in the low-lying Dutch delta. Cities like Dordrecht, Almere, Zwolle, and Rotterdam are currently rethinking their local flood risk and spatial planning strategies, because both the occurrence and the impact of high water will increase as a result of the expected impacts of climate change, combined with the ongoing spatial development of urban waterfronts.

In Rotterdam, the city government launched a research project in 2010, co-funded by Knowledge for Climate (KfC), to rethink flood risk approaches for unembanked areas. The current strategy of formal, mandatory elevation of ground levels in new spatial developments will not suffice and may be too expensive in the long run. For the subareas of Kop van Feijenoord and Noordereiland, both within the city district of Feijenoord, this research project (Van Veelen, 2013) provided two main local and integrated strategies for flood risk management. The first strategy is a collective-preventive strategy, aimed at building a flood defence wall or levee to protect the entire unembanked area. The second strategy is an individual-adaptive strategy that relies on the principle of the individual responsibility of real estate owners and private homeowners (residents) to protect their property in unembanked areas, e.g. by taking dry-proofing or wet-proofing measures.

As part of a follow-up of the 2010 KfC research project, the city decided in 2012 to start a co-creative research process to find out how these strategies could be refined in such a way as to make them implementable. Part of the research question was which governance arrangements could be designed to enable the implementation of these strategies.

Research objectives and research question

Previous KfC research projects (Veerbeek *et al.*, 2012; Nabaliek-Kronberger *et al.*, 2013) on potential physical-spatial measures to reduce the negative impacts of climate change indicated that an integrated local(ized) strategy could be cost effective. Also, a positive contribution could be made to the spatial quality of Feijenoord by coupling flood risk investments with spatial development and public investments (cf. Ward *et al.*, 2013). In various flood-prone areas of the City of Rotterdam (in addition to Feijenoord, e.g. Heijplaat), investing actors (such as housing corporations and real estate investors) and district authorities have an interest in developing alternative local adaptation strategies, including governance arrangements. However, the perceived lack in knowledge about financial (fiscal), legislative, and organizational instruments for sustainable and robust integration of flood risk measures with spatial development complicates decision making. This severely hampers the execution of decisions that are acceptable for local actors, because decisions made now will have their impact in the long term through investments in real estate and infrastructure (Mees *et al.*, 2012).

The major challenges for the parties concerned are complex. First of all, they must reach an agreement on the mutually accepted division of costs and benefits in the short and in the long term. Second, they will have to develop workable arrangements that secure responsibilities, management of risks, and long-term investments. Without innovative arrangements, private stakeholders (investors and real estate owners) will inescapably resort to traditional agreements that are unlikely to support sufficiently the local, integrated adaptation strategy.

The objective of our research is twofold: first, to develop knowledge about feasible forms of collaboration and arrangements to support an integrated localized flood risk strategy in the district of Feijenoord; second, to develop a generic method for synchronizing the physical-spatial measures (technical) with financial, legislative, and organizational aspects of strategies to reduce flood risks. The multiple objectives of our research efforts are framed in the following central research question: Which financial, legislative, and organizational arrangements can be used to develop and facilitate an integrated localized strategy for flood risk management?

To achieve this twofold objective, we applied an iterative and action-based method through which the generated knowledge can be applied in the real and tangible context of Feijenoord. Subsequently, we used our experiences and insights to develop more generic knowledge on how to construct arrangements for adaptation strategies in other contexts. The action-based research approach aims to organize a process of knowledge co-creation in an interdisciplinary team of researchers (from Erasmus University, Deltares, and Utrecht University), stakeholders, and policymakers from the City of Rotterdam and the district of Feijenoord, in order to develop feasible and acceptable governance arrangements to implement an integrated flood risk strategy. The co-creation process was built upon workshops with stakeholders and experts, literature study, and in-depth interviews.

In this chapter, we describe in detail the process of knowledge co-creation, the main problems, and the lessons we can draw from this case study. In particular, we shed light on the question of why action research in this case was perceived as a useful approach, the main difficulty that it faced in creating commitment among the intended participants, and our attempts to manage the process in order to deliver both policy-relevant knowledge and scientifically valid insights.

Climate adaptation in the unembanked areas of Rotterdam: a complex challenge

The region of Rotterdam is vulnerable to both tidal and pluvial floods. The majority of this urbanized region is protected by a network of primary flood defences that protect not only the city, but also a large part of the Randstad conurbation and the urbanized part of IJsselmonde Island. The region also has a large and highly developed urban area, which lies outside of the protection of the primary defence system. In the Rotterdam-Dordrecht floodplain about 65,000 people (distributed over 46 municipalities) live in these unprotected areas (Veerbeek *et al.*, 2010). The Rotterdam port industrial complex, which is vitally important for the Dutch economy and that of neighbouring countries, is also located outside the primary defence system. The unembanked area of Rotterdam, the focus area of this research, is a particularly relevant area for studying climate-adaptive development of unembanked urban areas because it is one of the most densely populated delta areas in the world.

Although these unembanked areas benefit from protection by the Maeslant Barrier, there is still a considerable risk of flooding (Veerbeek *et al.*, 2010; Van Veelen *et al.*, 2010). A major part of the area has been raised between 3 and 3.5 metres above average sea level. Only a few areas, like the Noordereiland and Heijplaat, run a high risk (e.g. between a yearly to a one every 100 years event) of sustaining flood damage (Veerbeek *et al.*, 2010). In the coming decades, the city will face two important developments. Although climate change increases the risk of flooding, the land use in the city centre and the former port areas outside the primary water defences is intensifying (Gemeente Rotterdam *et al.*, 2007). This aggravates the risk of future disasters, while at the same time the increased economic value and activities could cause the possible consequences of flooding to become more severe.

Flood characteristics of Feijenoord and Noordereiland

Although the Noordereiland and Kop van Feijenoord are both low-lying flood-prone areas, they differ in relation to characteristics such as flood frequency, water depth, and flood duration (Veerbeek *et al.*, 2012). The Noordereiland is a low-lying mound-shaped island that has to deal with high flood frequencies. The quays of the island are flooded at a yearly or 10-year flood event. At a 50-year flood event (3.04m+ above sea level), water can enter the basements and ground floors of buildings situated at the southern and northern end of the island. Because of the island's mound shape, the duration of a flood event is expected to be short. The higher part of the island also forms a relatively safe 'backbone' that can serve as an evacuation route when the low-lying areas along the quays are flooded.

The Kop van Feijenoord is a deep basin with a high risk of flooding (Nabielek-Kronberger *et al.*, 2013). In contrast to Noordereiland, the area can be compared to a bath tub that retains floodwater after a flood event. At a 50-year flood event (3.04m+ above sea level), half of the case study area would be flooded to a water depth of 50–75 cm. Water enters the ground floors of more than half of the buildings in this area. During extreme flood events, the exposed area hardly changes, but the water depths rise considerably to 80–100 cm, and serious damage to the façade and the interior of buildings can be expected (Veerbeek *et al.*, 2012). Due to the bath tub-like shape of the area, the floodwater cannot run off or drain to the river. It is expected that recovery in this area will take a couple of days.

Current flood risk management strategy

At this moment, there is no integrated flood risk policy for flood protection in the unembanked areas. The City of Rotterdam's prevailing long-term flood risk policy is based upon a formal regulation to raise the ground level of new building lots to the 1/10,000 storm surge flood level. The current storm surge flood height is set to a level that fluctuates between 3.90 to 4.10 m above sea level, depending on certain local conditions, such as wind direction and wave setup. This policy implies that new buildings and assets have to be raised to approximately 1 m above average street level. For existing urban areas, there is no additional policy or regulation in place to minimize the effects of a potential flood (Van Veelen *et al.*, 2010). Homeowners are responsible for possible damage caused to their homes by floods and must take precautionary measures, although currently they are poorly informed about local flood risks. Community disaster management is currently limited to closing-off quay sections and public areas. In addition, flood risk is not covered by home insurance.

The national government has delegated responsibility for flood risk in the unembanked area to local governments. Local authorities are responsible for deciding whether and under what conditions spatial development in flood-plain zones is allowed. Integrating flood risk management in spatial planning, however, has proved to be problematic. In the Dutch prevention-based flood risk management system, there is little or no experience with flooding, with the result that there is a lack of knowledge on flood-proofing measures and methods (Van Vliet, 2012; Van Vliet and Aerts, in press; de Moel *et al.*, 2013). As there is relatively little experience with flood-proofing measures, national laws and regulations are often not clear on their use and impact. Currently, flood risk management is not included in zoning plans, and risks are mentioned only on a voluntary basis in zoning documents to inform stakeholders. Flood-proof building regulations are not included in the National Building Act either, or in local building codes. In addition, flood zoning as an instrument in existing areas does not suffice, as land-use zoning plans are not appropriate legal instruments to change current functions.

Flood risk management for Feijenoord: towards alternative strategies

The mandatory elevation strategy has worked well in the recent past, when disused port areas were redeveloped within large-scale redevelopment projects, based on public land-development models and supported by public investments. The current policy of elevation of building lots and public land as an instrument to increase flood safety in a neighbourhood is a clear example of traditional hierarchical government steering (Kokx, 2012). Investments in elevating building plots and public space were developed within an integral land development plan and funded from the surplus value of the area after development. Consequent to the financial crisis and structural decline in demand for housing, business premises, and office space, the coming period will be characterized by limited need to develop new urban areas. It is expected that large-scale urban area developments will increasingly give way to small-scale transformations of the existing city, with other stakeholders and limited public funding involved, and plan periods that are prolonged or kept open (Krabben and Jacobs, 2012). These changes also affect urban flood risk policies. Elevating small-scale building plots is expensive and has a negative impact on the attractiveness of public space because of height differences between the first floor level and street level. Moreover, raising only new buildings to projected storm surge levels does not reduce risk to existing communities or of flooding of utilities and infrastructure.

The policy of elevation of ground levels generates important negative effects on other policy ambitions for the area, such as improving spatial and social quality at street level. This policy leads to buildings having different ground floor levels, and to creating blind corners and spaces that cannot be easily monitored and maintained. Furthermore, it leads to a discrepancy with respect to flood safety in the neighbourhood, because only new building locations will be elevated. From a social justice perspective, this approach can be questioned with respect to social equity: residents in social housing with fewer resources may be excluded from flood safety policies, whereas more affluent households in the new dwellings are included. Ultimately, the lack of an integrated flood risk policy, based upon incremental urban development processes and a strong separation of responsibilities between levels of government, contributes to an increased vulnerability of the area.

Integrated adaptive strategies

There is, however, a much broader spectrum of flood risk reducing measures, ranging from purely organizational measures such as risk communication to technical measures such as embankments and flood-proofing of existing buildings (Nabielek-Kronberger *et al.*, 2013; Van de Ven *et al.*, 2009). These flood-risk reducing measures can be applied both at the level of individual houses and at district and city level. In order to document the different measures, the City of Rotterdam has conducted research on adaptive strategies (Van Veelen, 2013). In this research project, the local measures that can reduce the vulnerability of the area and the effective use of area development processes have been identified and assessed on spatial quality, technical aspects, and financial feasibility. One conclusion from this research is that there are roughly two strategies to reduce flood risks in both subareas of Feijenoord district: a preventive community-level-based strategy and a resilient individual-level-based strategy. The first strategy is a collective-preventive one based on keeping water out of the area by gradually raising the low-lying quay areas to the projected level. Because of the specific morphological situation of the Kop van Feijenoord subarea, where a large part of the quay area has already been elevated, floodwater can be prevented from entering the area by raising a limited section of quays. This strategy seems less appropriate for Noordereiland because it would require the elevation of all quays around the mound-shaped island. This would be very expensive and destroy the unique cityscape.

The second strategy is an individual-adaptive one, based on creating a waterproof urban area that is able to recover within a reasonable amount of time from a flood event. This strategy is based on both dry-proofing and wet-proofing new and existing buildings, utilities, and civic infrastructure. As the Kop van Feijenoord area housing stock consists mainly of poorly renovated nineteen-century apartment blocks, flood-proofing the area implies retrofitting measures, which are in many cases technically and financially unfeasible. Flood-proofing new buildings, however, is much easier. The long-term feasibility of this strategy will depend on the social housing corporation's future development strategy (renovation or rebuilding). For Noordereiland, this strategy would be more feasible because floods are likely to be of short duration given the island's mound shape. It could make sense for private homeowners to take dry-proofing and/or wet-proofing measures to protect their houses in short periods of flooding.

The proposed solutions for Noordereiland and Kop van Feijenoord differ from the current flood-safety policy in the unembanked areas, in more than just the technical domain. They also call for new types of partnerships between public institutions and private parties such as homeowners, housing associations, and real estate investors. Private parties are jointly responsible for the construction, financing, and management of the proposed adaptive flood measures. This shared responsibility for local flood management creates a host of new issues, such as the question of how the short-term costs of flood protection can be compensated by the long-term benefit. Might it be possible to create a local flood protection fund to be responsible for the costs and benefits of flood management?

Methodological approach: action research to design governance arrangements

To answer the central research question, we developed an action-oriented research approach. The research project was organized with the objective of facilitating the step from expert-led to expert-fed research. Within a timespan of approximately six months, stakeholders were actively involved in discussing, digesting, scrutinizing, and modifying expert knowledge about potential physical-spatial measures to accommodate the impacts of climate changes in the Feijenoord area. This expert knowledge had been generated in earlier KfC projects (Van Veelen *et al.*, 2010; Van Veelen, 2013).

Research approach: action research

Our research approach is action oriented because we actively involved stakeholders in exploring new insights about feasible arrangements for alternative strategies for flood risk management. According to Friedman (2001: 160) 'action science attempts to bridge the gap between social research and social practice by building theories which explain social phenomena, inform practice, and adhere to the fundamental criteria of science.' Action research is driven by reflection on interventions aimed at changing certain situations (cf. Argyris *et al.*, 1985). Action research is specifically suitable for initiating and guiding change processes. By first examining what is going on, and subsequently changing the undesired situation (e.g. a problem or missed opportunity) in small steps, under continuous (reflexive) monitoring, a bridge is built between 'is' and 'ought,' according to those involved (cf. Biggs, 1999). In that respect, action research is a productive method to give shape to more reflective practice, in and between organizations, through direct collaboration between practitioners and researchers (cf. Duijn *et al.*, 2010). Our approach was designed as an iterative interaction process between three groups: a stakeholder group, an expert group, and a so-called core group, including a facilitator who had to organize the iterative interaction between these groups. In terms of Chapter 2 of this book, the depth of interaction can be qualified as co-production, and our level of involvement in the policy process as intervention. It can be seen as a mix of cooperative inquiry and participatory action research, as it combines elements of co-production (whereby the division of roles became blurred), and participation, as we tried to empower local stakeholders to co-design local implementation arrangements.

Our approach is based on the following general theory of action: A structured interaction process between representatives of stakeholder organizations and experts, guided by a core group (including researchers and facilitator), will support the formulation of governance arrangements for local adaptation strategies that can create a foundation for tangible agreements between stakeholders who are active in the development of unembanked areas in general, and more specifically in Kop van Feijenoord and Noordereiland.

Iterative interaction as designed vs. iterative interaction as implemented

The implementation of a research project in a live situation, involving actual stakeholders and using tangible problems, is likely to undergo adjustments and changes. We had planned to organize five workshops to facilitate the stakeholder–expert interaction, all in the presence of the core group and aimed at facilitating two meetings with local residents[1] prior to and after the stakeholder–expert workshops. For the stakeholder–expert interaction, we intended to organize two separate workshops with stakeholders and two with experts, followed by a joint final workshop to conclude the iterative process. First, stakeholders were to separately discuss the research objective and case study area. Second, the same was to be done by the experts. Third, stakeholders were to be asked to formulate the principles for the intended arrangements for local adaptation strategies. In turn, in a fourth workshop experts would review these principles and provide suggestions for improvement or adjustment. Fifth and last, stakeholders and experts would collaboratively discuss and finalize remaining issues around the desired arrangements. However, this design for the interaction process had to be adjusted shortly after its start for the following reasons.

First, the Planteam Feijenoord – a joint venture between public authorities and private property developers – was dismantled only a few weeks after the project started; this meant a major setback for the project because it had been planned that this team would be the key supplier for the core team. More importantly however, this team would have been in a position to make the desired agreements with other stakeholders on new local adaptation strategies and implement alternative measures. The dismantling of the Planteam resulted from the conclusion that no large-scale developments in the Feijenoord district would be initiated in the short term because of the financial crisis. There was no clear successor or contact person to build a new core group, so we had to start all over again to find a productive connection within the Rotterdam municipality. This was done by organizing a initial meeting with relevant civil servants from the municipality and from the district of Feijenoord. Involving the district authority was not easy, as climate change adaptation does not come within its primary remit. The upcoming reorganization of the district governments in Rotterdam was another obstacle to active participation by Feijenoord's district representatives with regard to tasks that are not part of their core business, such as flood prevention. The same occurred when we tried to make connections with two key actors in Feijenoord area development, housing corporation Woonstad and real estate investor AM Wonen. It appeared difficult to involve these parties in research on climate adaptive area development. The knowledgeable professionals in these organizations had only limited time and willingness to participate. Lastly, the disintegration of the Planteam and the slowing down of Feijenoord area development turned the research process into a more political issue as it became a stage on which the developing parties could display their interests. It was not intended that the initial stage of the research project would be used to negotiate about the distribution of costs and benefits, but attempts were made immediately to use it for this purpose.

Second, during the research process, we encountered limited availability of both administrators and civil servants, caused by what is often referred to as administrative overload. This overload had several causes. Institutional uncertainty about the future relation between city and districts was caused by the upcoming (in 2014) reorganization (or partial abolishment) of the city districts. This uncertainty led to ambiguity about the role of professionals from the city and from the district in the project. In addition, at the time, many other exploratory studies were being implemented with regard to finding ways to climate-proof vulnerable parts in the Rijnmond region and in Rotterdam, especially Feijenoord. Many of the same professionals (e.g. from city and district authorities) were asked to participate in many different projects, causing them to be ever more selective about devoting time to these studies. Much effort was needed to convince them to participate, and in some cases they decided to quit. Lastly, the planning process for Feijenoord had its own dynamics and evolved rather autonomously from the research project. Attempts to connect planning and research closer to each other failed because of the dismantling of the Planteam (see above). Professionals refused to devote more time to the project because of its limited immediate interest for them and because their time was claimed by their administrators.

Third, we experienced difficulties in involving citizens and homeowners in the research project. The lack of a sense of urgency and the limited level of cooperation stood in the way of involving them at short notice. A meeting with some members of the core group revealed that administrators were reluctant to agree to involve citizens and homeowners in this stage of the policymaking process with regard to the city's adaptation strategy. Therefore, the intended interaction with citizens and homeowners was to be framed as 'doing collaborative research' instead of allowing for a formal point of contact for questions about policies or measures with regard to local adaptation strategies. In the initial meeting with professionals from the city and the district of Feijenoord, it was concluded that there was a significant difference in real estate ownership in both subareas, Noordereiland and Kop van Feijenoord. Consequently, it was decided that a separate research process would be initiated for each subarea, aimed at organizing interaction with the key stakeholders, i.e. the real estate owners. This decision was grounded in the ambition to direct the research efforts at stakeholders who were authorized and competent to change the existing practice of climate adaptation, that is, to prevent damage and nuisance caused by flooding (or high water levels). In Kop van Feijenoord, Woonstad and AM Wonen are the key stakeholders because they own the largest amounts of real estate and building land for new urban development. Thus, these parties were in the position to change the current practice of urban development, for instance by renovating existing property or building climate-proof houses. Collaboration with the city would be necessary because of its responsibility to adjust the public space and infrastructure in the subarea accordingly. In the subarea of Noordereiland – mainly the western part – the key stakeholders were private homeowners because they were responsible for taking measures against damage and nuisance caused by high water levels.

The conclusion in relation to the great difference in homeownership between Kop van Feijenoord and Noordereiland, and therefore in authorization and responsibility to act, led to the decision to attempt to involve citizens and homeowners only in the Noordereiland subarea. Therefore, the intention to organize a meeting with local residents from both subareas at the start and at the end of this research project was abandoned.

In the end, the action research project – that is, the organized interaction between stakeholders and experts in workshops – was carried out between November 2012 and June 2013 (see Figure 10.1). The participating stakeholders

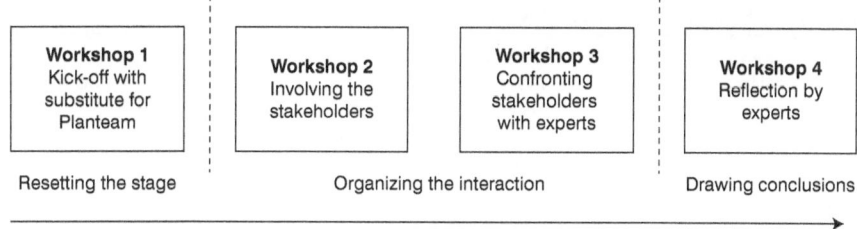

Figure 10.1 Scheme of the executed action research process

were representatives of the City of Rotterdam (several departments), the district of Feijenoord, real estate investor AM Wonen, housing corporation Woonstad, and private-sector firm Hunter Douglas. Experts from Utrecht University, RebelGroup, Rotterdam Community Solutions, and Deltares were involved to reflect on the stakeholders' suggestions. The substantial contribution of both stakeholders and experts to the research project was as intended (see above). The main difference to the intended action research process was the change in the number of workshops (four instead of five) and early interaction between both groups in one workshop. In addition, we concluded our research efforts by broadening the set of experts in the latter workshop to reflect on the provisional results from the perspective of the Delta Programme. This was done because the responsibility for flood risk management of unembanked areas is (still) not properly settled.

The researchers prepared and facilitated the workshops, and processed and reported the results back to the participating stakeholders and experts. As indicated above, they had to take continuous decisions to keep the research process on track.

Substantial results

The outcomes of this action research project can be divided into three parts:

1   functional requirements for a future governance arrangement (based upon interviews with key stakeholders);
2   tentative design of a possible governance arrangement that meets these requirements (based upon interactive sessions with stakeholders and experts);
3   necessary institutional adjustments in the current institutional regime (based upon reflection by involved experts).

Functional requirements

Our interviews with key stakeholders unearthed issues that have to be taken into account in relation to designing a new governance arrangement. We call these issues *functional requirements* (see Table 10.1). The new governance arrangement must take these requirements into account and provide these *functionalities*.

Table 10.1 Functional requirements for an adaptive governance arrangement

FR1	Communication/awareness	How to create a sense of urgency/how to increase awareness about the increased risk of flooding and the necessity to take measures? *The arrangement has to provoke more commitment and dedication from residents.*
FR2	Role of municipality	What role can the municipality fulfil in facilitating adaptive flood risk strategies, without taking over other actors' responsibilities? *The city government is willing to fulfil a facilitating role but is very cautious about accepting new responsibilities.*
FR3	Appropriate financing	How to oblige or tempt actors to pay for flood risk measures? How to pool resources and enable the synchronization of investment paths? *The arrangement has to enable collaborative financing of flood measures and their maintenance.*
FR4	Legitimacy	What is the reach of existing formal competencies and what can be done within these competencies? *The arrangement has to recognize and accommodate the existing distribution of responsibilities.*
FR5	Compliance	How to prevent moral hazard and free-riding? *The arrangement has to prevent moral hazard and free-riding and has to secure cohesion and accomplishment.*
FR6	Joint responsibility	How to make flood risk management a joint responsibility, a shared enterprise of public, private, and societal actors? *The arrangement has to facilitate collaboration and sharing responsibilities.*
FR7	Connecting long-term challenges and short-term ambitions	How to connect short-term (economic, spatial) developments to the long-term ambition of making unembanked areas climate-proof? *The arrangement has to facilitate collaboration on both short- and long-term ambitions and to manage a flexible adaptation pathway.*

These functional requirements were derived from interviews with 10 key stakeholders and four workshops with the various stakeholders we could mobilize to join this research project. The requirements were formulated in order to function as design principles for an adaptive governance arrangement.

Tentative design of an adaptive governance arrangement

Due to the particular characteristics of the existing governance constellation, the governance arrangement has to be the joint responsibility of many actors. Land- and homeowners are themselves responsible for dealing with flood risk, but the city government is responsible for risk communication and feels a moral obligation to take mitigating measures when risks seem to be increasing. Only when these various responsibilities are pooled and combined can an adaptive flood risk strategy be implemented (cf. Van den Berg *et al.*, 2013).

The collective-preventive strategy in particular requires a governance arrangement that enables collaboration between a wide variety of both public and private actors. This strategy could be perceived as the sole responsibility of the municipality. However, we advocate collaboration between public and private actors in conceiving and implementing a feasible governance arrangement. In our view, it would be undesirable if investments were made solely by public actors but some of the revenues accrued to private stakeholders. Such a collaborative, public–private arrangement has at least four components:

1   a financial component that allows for sharing rather different budgets and contributions (FR3);
2   an organizational component that facilitates collaboration and enables implementation (FR6);
3   a substantial component in which both long-term and short-term ambitions are integrated in a collaboratively defined vision (FR1, FR7);
4   a legal component with agreements about distribution of roles and responsibilities (FR2, FR4, FR5).

1: Area fund

A collective financial arrangement (like an area fund) is necessary to collect the various contributions of stakeholders (residents, companies) and to bring together different public budgets in order to enable the financing of collective provisions. Such a fund construction makes it possible not only to pool resources, but also to overcome differences in tempo and timing of different investments. A fund can be used as a savings account. It can also function to prefund some investments and to keep the money available (which is not possible within public budget schemes). A fund makes it possible to remove the restrictive conditions of public budgets in respect of their timeframe and allocations. A fund facilitates collaboration and co-production by enabling co-financing. Many stakeholders deemed it legitimate for the water board also to contribute to this fund because they also levy taxes for flood safety investments in unembanked areas.

2: Area authority

An inter-organizational structure (either more informal or more formal) is necessary to facilitate collaboration between the various actors. Such an organizational structure has to have the capacity to facilitate both the preparation and implementation of the area vision and the agreements from the area contract (see next section).

The area authority does not necessarily have to be a new entity with new competencies. Rather, it will function as an auxiliary arrangement that makes it possible to collaborate and to join forces while keeping the existing distribution of competencies unchanged. Such a management position is necessary because structural cooperation between various actors is necessary in order to align public and private developments relating to spatial investments and maintenance activities. The authority is the embodiment of the actors' (investors, residents, and public authorities) joint responsibility to realize a more adaptive flood risk strategy in an intelligent way, based upon combining/synchronizing small and large actions, taken by different actors.

3: Area vision

A collaborative long-term vision, allowing consideration of short-term urgent issues, is indispensable. A long-term vision is necessary because it will take decades to realize a specific adaptation strategy. Short-term visions are necessary to align current investments and developments to this long-term ambition, and to take into account new circumstances and new hot topics.

Maybe the process of coming to a shared vision is even more important than the vision itself. The process can be seen as an attempt to realize joint image-building, to create awareness and commitment. Thus, it is an investment in communication and building broad support. It invites other actors to think about their own future ambitions, to clarify what they can contribute to, and expect from, a flood risk strategy at junctures that best fit their own agenda.

4: Area contract

It is necessary to give such a vision a formal, obligatory status and to make agreements binding on all participants. A formal public–private agreement guarantees commitment and can function as a means to enforce accomplishment when actors try to refrain from contributing to the collective strategy. An area contract translates the agreements made in the area vision into concrete, binding covenants that specify which actor is responsible for which task/contribution at which time. These explicit arrangements also contribute to the various actors' awareness of their own responsibility.

Institutional design requirements

Finally, the action research project also addressed two issues relating to 'institutional design': institutional or legal adjustments necessary to enable alternative flood risk strategies.

1   The role and responsibilities of the regional water board and Public Safety Region (regional disaster management authority). Although the water board has the knowledge and the capacity to organize flood risk safety, it has no responsibility under national law (*Waterwet*). By changing the current distribution of responsibilities it becomes possible for water boards

to collaborate in new, adaptive flood risk strategies. Surprisingly, the local residents pay (reduced) taxes to the water board for water system management. However, in the event of high flood levels, the water board has no responsibility for prevention as the area is unembanked.

It is also important to reconsider the role of Public Safety Region. This agency has a responsibility for risk and disaster management, but its involvement in spatial planning processes is quite low. When a more adaptive strategy is adopted, the involvement of Public Safety Region will be important to manage evacuation and assistance, for example by organizing disaster exercises and warning about risks.

2   The role and responsibility of the national government – especially in the long run. The flood risk safety in unembanked areas is highly dependent upon decisions taken by the national government with regard to the main water system (river discharge distribution, closing regime of the Maeslantkering flood barrier). These decisions (even to maintain the current regime, despite worsening climate impacts) directly affect flood risk safety in unembanked areas. This implies that it is legitimate to reconsider the responsibility of the national government for unembanked areas if climate change persists. Municipalities need to take a more pro-active role in developing local flood risk management strategies to ensure a better integration of flood risk management in local development and management processes. Moreover, to date, the Delta Programme has not taken responsibility for making policies and/or taking measures to protect unembanked urban areas, leaving it entirely to the municipalities. These organizations are not appropriately equipped to take on this task. National government needs to decide either to take up the challenge itself or to equip the municipalities to do so.

Reflection: lessons and insights about how to use action research

Reflection on the changed approach suggests that the available time was too limited to realize the pre-set ambitions of implementing action research in this context. Reflecting on the general theory of action formulated earlier leads to the following observations.

Organizing the engagement of professional stakeholders in structured interaction with experts proved to be difficult and very time-consuming. During the research project, stakeholders felt (too) little sense of urgency to overcome the attitude of 'let's wait and see what the municipality plans to do, and then we may decide to collaborate'. This attitude indicated that climate adaptive urban development in Kop van Feijenoord was (and is still) not regarded as a collective responsibility and interest. The ambiguity for citizens and homeowners with regard to the question of to whom they should appeal in high water situations did not help either. The Feijenoord district authority is the nearest government organization for them but has no formal responsibility, policy, or resources to deal with flood situations. The fact that the financial-economic driver for urban development had vanished did not help either, especially when this induced the municipality to shelve the masterplan process.

However, our action research project was successful because it resulted in a governance framework that can help develop a tailor-made arrangement for Feijenoord. Although our research project was not capable of steering the local flood risk strategy towards implementation, the outcomes of our study have been used to raise awareness of the importance of governance issues for adaptive flood risk management in several political discussions at local and regional level. Moreover, the proposed adaptive governance arrangement will be included in advice on local flood risk management in Feijenoord's revised redevelopment strategy.

Implications for the study of climate change adaptation

Our case study underlines the importance of creating awareness when adaptation strategies are being developed (Hulme, 2009). Local people stick to the traditional distribution of roles and responsibilities and do not feel the urgency to invest in the uncertain consequences of climate change. Creating a sense of urgency requires thoughtful communication (Moser and Dilling, 2004) and this cannot take place without processes of social learning and frame reflection (Dewulf, 2013).

Furthermore, our case study research also highlights the importance of analysing the current institutional regime and its impact upon the governance of climate issues. Sometimes, institutional choices from the past hinder the implementation of effective adaptation strategies, and thus it may be necessary to redesign existing regimes (Tompkins and Adger, 2004).

Finally, and related to the preceding issue, we have to know more about the 'design space' in the case of climate change adaptation. From our case, we can conclude that the willingness to change existing institutional arrangements is very limited; this also makes it rather difficult to adjust the current regime. When the design space is too limited, it is necessary to think about how to construct auxiliary arrangements that enable innovative solutions without changing the institutional regime.

Implications for the governance of climate change adaptation

The lack of urgency and commitment and the limited sense of collective responsibility to deal with the challenges of flood-proofing already developed urban unembanked areas leads to the conclusion that these areas are treated as a 'hot potato' that is tossed around between the actors involved. Formally, the municipality has the responsibility but not the resources and expertise to take up this challenge. Water authorities such as water boards and Rijkswaterstaat do have the resources and expertise but are not responsible. Even the national Delta Programme – aimed at revising Dutch flood policy – excludes these areas from its objective, measures, and resources. Therefore, the development of these attractive, but vulnerable, areas is stuck in deadlock. From our analysis, we conclude that the only effective way to bite the bullet is to redistribute responsibilities.

Implications for conducting action research

Our case study shows that it is very difficult to organize action research for an issue that is deemed not very urgent. The participation of citizens and homeowners was very limited, and the first aim should be to create awareness about self-reliance and taking on responsibility. Subsequently, joint investigation must be aimed at what is needed – e.g. practical knowledge or resources – to take measures within their span of control. Such an approach requires a more extensive process than was possible here within the constraints (time, budget) of this research project. Also, it calls for a more formal administrative commitment and mandate for both the municipal and the district authority to work with local residents. After all, flood safety is a politically sensitive issue that should be governed by formally responsible administrators. In fact, had the democratic anchorage of our action research project been stronger, it would have had more impact (Edelenbos, 2005).

Our action research project was not aimed at organizing a process of social learning with stakeholders in order to develop a joint feeling of collective responsibility. With hindsight, we can conclude that such a learning process is necessary before it is possible to develop legitimate governance solutions and subsequent implementation arrangements (Collins and Evan, 2008).

Note

1 Resident involvement had to be limited because of the short timeframe in which the research project had to be carried out.

References

Argyris, C., R. Putnam, D. McLain Smith (1985) *Action research: concepts, methods, and skills for research and intervention*, San Francisco, CA: Jossey-Bass Publishers.

Berg, H. van den, M.W. van Buuren, M. Duijn, D. van der Lee, E. Tromp and P. van Veelen (2013) *Klimaat-adaptieve ontwikkeling van binnenstedelijk buitendijks gied: van hete aardappel naar zure appel? Governance van lokale adaptatiestrategieën: de casus Feijenoord*, KvK report 103/2013, Utrecht: Kennis voor Klimaat.

Biggs, J. (1999) *Teaching for quality learning at university*, Buckingham: Open University Press.

Collins, H. and R. Evans (2008) *Rethinking expertise*, Chicago, IL: University of Chicago Press.

Dewulf, A. (2013) Contrasting frames in policy debates on climate change adaptation, *Wiley Interdisciplinary Reviews: Climate Change*, 4(4): 321–330.

Duijn, M., M. Rijnveld and M.J. van Hulst (2010) Meeting in the middle: joining reflection and action in complex public sector projects, *Public Money and Management*, 30(4): 227–233.

Edelenbos, J. (2005) Institutional implications of interactive governance: insights from Dutch practice, *Governance*, 18(1): 111–134.

Friedman, V. (2001) Action science: creating communities of inquiry in communities of practice, in P. Reason and H. Bradbury (eds) *The handbook of action research*, 159–170, London: Sage.

Gemeente Rotterdam, Waterschap Hollandse Delta, Hoogheemraadschap van Schieland en de Krimpenerwaar and Hoogheemraadschap van Delfland (2007) *Waterplan 2 Rotterdam: werken aan water voor een aantrekkelijke stad*, Rotterdam: Gemeente Rotterdam.

Hulme, M. (2009) *Why we disagree about climate change*, Cambridge: Cambridge University Press.

Kokx, A. (2012) *Het vergroten van de adaptieve capaciteit in een buitendijkse binnenstedelijke gebiedsontwikkeling: een draagvlakverkenning*, KvK report 89/2013, Annex D, Utrecht: Kennis voor Klimaat.

Krabben, E. van der and H.M. Jacobs (2012) Public land development as a strategic tool for redevelopment: reflections on the Dutch experience, *Journal of Land Use Policy*, 30(1): 774–783.

Mees, H.L., P.P. Driessen and H.A. Runhaar (2012) Exploring the scope of public and private responsibilities for climate adaptation, *Journal of Environmental Policy & Planning*, 14(3): 305–330.

Moel, H. de, M. van Vliet and J.C. Aerts (2013) Evaluating the effect of flood damage-reducing measures: a case study of the unembanked area of Rotterdam, the Netherlands, *Regional Environmental Change*, doi:10.1007/s10113-013-0420-z.

Moser, S.C. and L. Dilling (2004) Making climate hot: communicating the urgency and challenge of global climate change, *Environment*, 46(10): 32–46.

Nabielek-Kronberger, P., D. Doepel and K. Stone (2013) *Design research adaptive strategies in the unembanked area of Rotterdam*, KvK report 89/2013C, Utrecht: Kennis voor Klimaat.

Tompkins, E.L. and W. Adger (2004) Does adaptive management of natural resources enhance resilience to climate change? *Ecology and Society*, 9(2): 10.

Veelen, P. van (2013) *Adaptive strategies for the Rotterdam unembanked area: synthesis report*, KvK report 89/2013, Utrecht: Kennis voor Klimaat.

Veelen, P. van, H. Meyer, E. Tromp, S. Plantinga and K. Batterbee (2010) *Klaar voor hoogwater: verkennend onderzoek naar adaptieve strategieën voor het buitendijks gebied in de hotspot Rotterdam*, KvK report 025/2010, Utrecht: Kennis voor Klimaat.

Veerbeek, W., C. Zevenbergen and B. Gersonius (2010) *Flood risk in unembanked areas: part C vulnerability assessment based on direct flood damages*, KvK report 022C/2010, Utrecht: Kennis voor Klimaat.

Veerbeek, W., R. Ashley, C. Zevenbergen, J. Rijke and B. Gersonius (2012) Building adaptive capacity for flood proofing in urban areas through synergistic interventions, in *Proceedings of the 7th International Conference on water sensitive urban design*, 21–23, Melbourne: Centre for Water Sensitive Cities, Monash University.

Ven, F. van de, E. Luyendijk, M. de Gunst, E. Tromp, M. Schilt, L. Krol, B. Gersonius, C. Vlaming, L. Valkenburg and R. Peeters (2009) *Waterrobuust bouwen: de kracht van kwetsbaarheid in een duurzaam ontwerp*, Rotterdam: Beter Bouw- en Woonrijp Maken/ SBR.

Vliet, M. van (2012) *Deelrapport ruimtelijke ordening en bouwvoorschriften: juridische haalbaarheid van maatregelen Kop van Feijenoord*, KvK report 51/2012, Utrecht: Kennis voor Klimaat.

Vliet, M. van and C.J.H. Aerts (in press) Adaptation to climate change in urban water management: flood management in the Rotterdam Rijnmond Area, in R.Q. Grafton, K.A. Daniell and C. Nauges (eds) *Understanding and managing urban water in transition*, New York: Springer.

Ward, P.J., W.P. Pauw, M.W. van Buuren and M.A. Marfai (2013) Governance of flood risk management in a time of climate change: the cases of Jakarta and Rotterdam, *Environmental Politics*, 22(3): 518–536.

11 Action research for the governance of adaptation to climate change

Conclusions and lessons learnt

Mathijs van Vliet, Arwin van Buuren and Jasper Eshuis

Introduction

The governance of adaptation to climate change is beset by many complex challenges. As it is a relatively new policy terrain, policy agendas, instruments, and arrangements are still under construction. This implies a strong perceived need for application-oriented knowledge, as well as for policy advice on how to govern climate change adaptation and deal with the specific challenges that it poses. In other words, applicable knowledge is needed on how to govern adaptation to climate change. As discussed in the first chapter, action research methods seem promising to realize this ambition.

This book therefore aims to further the application of action-oriented research approaches in general, and in the domain of climate change adaptation in particular. It aims to do so by learning from applications in different institutional contexts and with different methods. In the previous chapters, different approaches to action research in the field of governance of adaptation to climate change are presented, describing how action research works in such a complex and relatively new context, and highlighting potentials and pitfalls for action-oriented research approaches in this emerging domain.

Chapter 1 started by sketching the challenges of climate adaptation governance and why an action research approach could potentially contribute to dealing with these challenges. Chapter 2 continued by describing action research and its many faces in more detail. These two chapters thus sketched the societal and theoretical context for the various case studies as presented in Chapters 3 to 10 of the book. These case-oriented chapters provided a wide variety of cases with various action research approaches, various levels of stakeholder/actor involvement, and different aims and research questions.

Building on the rich information in the case study chapters, this final chapter aims to provide an overall reflection on the use of action research in the field of governance of climate change adaptation, on the basis of three research questions:

- What forms of action research are applied in the context of governance of adaptation to climate change and how do they work?

- What are the main potentials and pitfalls for action-oriented research in this domain?
- What did our research teach us about the characteristics of governance of climate change adaptation?

This chapter looks at the types, levels, and methods of action research (described in Chapter 2) used in the various cases, and it reflects on the reasons behind these choices. The following section provides the reader with a short overview of the practices of action research, as presented in the previous chapters. What has been done in the various cases, what levels of intensity in collaboration and methods were used, and why?

The third section studies the scientific contribution of action research. What insights into the governance of climate change adaptation have the previous chapters provided? How can an action researcher safeguard scientific criteria such as recoverability and independence? The section finishes with a reflection on the normative issues that surround action research.

The chapter continues by discussing the practical value of action research in climate change adaptation processes in the fourth section. What is the contribution of action research to climate change adaptation governance practices? We highlight some of the practical issues with which the authors of the previous chapters were confronted, such as the time horizon and perspective of their research vis-à-vis policymakers' ongoing concerns.

The concluding section studies the balance between the pros and cons of action research for the governance of adaptation to climate change, and how action research can proceed in this field.

Practices of action research: what was done in the cases

Action research can take various forms, as illustrated in Chapters 3 to 10. Table 11.1 provides an overview of the different methods, levels, and types of action research used in this book. A couple of observations can be made based upon this overview table.

The empirical chapters reflect the fact that the field of adaptation to climate change is young and that most countries are still in the process of developing policies and strategies. Most chapters therefore aim to evaluate or study the on-going development of policy processes. Only Chapters 7 and 10 focus on the implementation of adaptation strategies.

Further, the empirical chapters are often based on a mix of traditional data collection methods (interviews, surveys, and document analysis) and more interactive methods (reflective workshops, focus groups, or games). In many cases, for example in Chapter 10 (van Buuren *et al.*), data collection is used as a preparation for interaction and knowledge co-production between scientists and practitioners. Chapters 3 and 5 (Vink *et al.* and Boezeman *et al.*) show how these data collection methods can also be used in an action research approach to increase reflexivity. Very intensive action research methods, such as researchers

Table 11.1 Overview of empirical chapters

Chapter	Action research level	Action research type	Action research depth	Main method(s)	Aim
3. Vink et al.	Participatory observation and reflection	Elements of action science and learning evaluation	Consultancy, co-decision	Partnering with a 'bad' guide, frame reflection between policymakers and scientists, cooperative data inquiry	Evaluate the potential of partnering with a 'bad' guide as an action research approach to dealing with the various views that complicate societal adaptation to a changing climate
4. Huntjens et al.	Observation and reflection	Action science	Consultation, co-decision	Group model building, multi-stakeholder dialogues, surveys	Develop a climate change adaptation strategy. Ensure that the priorities and challenges of actors at different governance levels are included. Study how action research methods can be applied in a non-Western culture
5. Boezeman et al.	Observation, reflection, and to some extent intervention	Elements of action science and learning evaluation	Information, consultation, co-decision	Honest broker, multiple cycles of observation, interviews, workshops, and (re)action to enhance reflection	Study how existing governance institutions deal with climate adaptation and participation as new requirements in knowledge production and policymaking. Facilitate a collaborative learning experience on participation
6. Roggema et al.	Participatory observation and reflection	Participatory action research	Consultation	Design charrette, interviews	Study the influence that social learning through charrettes may have in communities and the way this contributes to more adaptable spatial futures

7. Ellen et al.	Reflection	Action science	Consultation, co-decision	Serious gaming, interviews, desk studies, survey	Analyse the implementation of adaptation policy in practice and the role of flexible policy arrangements in particular. Facilitate the implementation of adaptation policy. Widen the scope of implementation measures taken into account
8. Driscoll and Lehmann	Reflection	Action science and action inquiry	Consultancy	Serious gaming, digital video, participant observation, interviews	Study effects of the combination of mitigation, adaptation, and sustainable urban development within the same gaming space. Develop and use games as a data collection tool
9. Schenk and Susskind	Observation and reflection	Action science and cooperative enquiry	Consultancy	Role-play simulation exercises, before-and-after surveys, recording game and debriefings	Encourage adaptation planning and explore the opportunities and challenges associated with different approaches, decision-support tools, and adaptation options
10. Van Buuren et al.	Reflection and intervention	Cooperative inquiry and participatory action research	Co-production	Knowledge co-creation, workshops, literature study, in-depth interviews	Develop knowledge about feasible forms of collaboration and arrangements to support a locally integrated flood risk strategy. Develop a generic method for synchronizing the (technical) physical–spatial measures with financial, legislative, and organizational aspects

working in a government institution for longer periods of time, are more difficult to arrange and implement and take even more time and resources, which are often lacking. Driscoll and Lehmann (Chapter 8), however, used on-site observations to gather enough insights to be able to develop games.

The difficulties of conducting intensive action research methods are reflected in the levels of action research applied in the cases, which were mainly (participatory) observation and reflection. More intensive levels of action research are often difficult to conduct as many organizations are not willing to allow experiments, especially if initiated by outsiders. The new and contested nature of climate change and adaptation might further decrease organizations' willingness to participate in experiments because governmental agencies are still thinking about their appropriate role and their subsequent strategies. In none of the cases has the highest level of action research – experimentation – been applied (although policy experiments are now and then being undertaken, see for instance McFadgen *et al.*, under review; Berkhout *et al.*, 2010). However, via the relative new action research tool of serious games, experimentation is possible outside the 'real world'. Intervention was possible in the case study described by van Buuren *et al.*, (Chapter 10), which took place in a project started and coordinated by the City of Rotterdam. It must be noted that, in some cases that were not labelled as interventions, reflexive practices such as games or workshops did have effects on the policy process and can thus be seen as interventions.

Table 11.1 also shows that in some cases action research was limited to reflection and participatory observation (e.g. Boezeman *et al.*, Chapter 5). This reflects the fact that some authors deliberately choose a more distanced and therefore more modest level of involvement in cases, as they do not want to become part of a potentially controversial policy process and lose their independence as researchers. Games are frequently used in the cases described. They have two main functions: to analyse decision makers and other actors' motives, strategies, and logics in specific situations, and to stimulate reflection and learning by confronting actors with new ideas and inviting them to use these ideas in simulated decision situations. It is a rather efficient research method, as actors can be observed well in processes that normally take considerably longer to play out. The rise in the use of games is reflected in the number of chapters dealing with games. As a relatively new method in the action research domain, we address them in more detail in various parts of this chapter: what are the implications of using games in action research?

The diversity of approaches and methods applied in the empirical chapters cannot be fully captured through the action research level and depth classifications in the table. In that sense, these classifications, developed in Chapter 2 of this book, clearly have their limitations. Reflecting upon this classification scheme, we could include distance from the ongoing governance process as an additional typology for action research in relation to governance.

Another limitation of the classification scheme is that it classifies, but does not give any information about the underlying motives for choosing a particular action research level. As Table 11.1 shows, almost all empirical research in this book aims to consult stakeholders and policymakers. However, the reasons for doing so are very different. Some researchers aim to study policy processes and institutions, others desire information on why some implementation measures are used more often than others. Often, practical aims (e.g. to assist with participatory processes; to develop climate change adaptation strategies) are combined with scientific or methodological aims (e.g. to develop methodological innovations; to study the use, in a non-Western culture, of action research methods developed in and for the West).

The scientific value of action research

Insights into the governance of climate change adaptation

The various cases presented in the previous chapters not only yielded insights into the methodologies of action research, but also provided very interesting insights into ongoing governance practices, contextual factors that matter, and strategies that work (or not). This corresponds to the second aim of this book, to gain more knowledge on the empirics of the governance of climate change adaptation.

Climate change adaptation is clearly a challenge dealt with in complex, multi-level networks of many different but interdependent actors. These actors have not only different perceptions on how to deal with climate change effectively, but also their own agenda relating to other ambitions like agriculture, transportation, or nature development. Governance of climate change adaptation essentially has to do with synchronizing the adaptation ambition with other agendas and ambitions. In other words, adaptation to climate change has especially to do with making creative connections with other agendas and with mainstreaming adaptation into existing policies with regard to water management, planning, and infrastructure (van Buuren *et al.*, 2014).

However – as also becomes clear from the various cases – there is much to be done to increase the sense of urgency about, and thus the political commitment to, climate adaptation, although there are significant differences. A lack of urgency and commitment is a serious hindrance to achieving effective results, as shown for instance in Chapter 10 (van Buuren *et al.*). The various games can play, and often have played, a valuable role in increasing awareness and commitment among policymakers.

However, that is not the whole story. The knowledge about policy arrangements that fit the ambition of adaptive strategies is rather limited (see Chapter 7, Ellen *et al.*). And even when there is enough urgency, commitment, and knowledge, the various cases underline the conclusion that climate adaptation has to deal with the danger of implementation gaps: there are many possible strategies, but the resources and competences to implement them are rather limited. This also relates to another observation. Although climate change adaptation shows many similarities with other wicked policy issues, the challenge is also unique in a number of respects. Its long-term and multi-scale character, its interconnectedness with a variety of issues, and the uncertainty and ambiguity of causes and consequences challenge existing routines (see Chapters 3, 4, 5, and 10). However, there is much conservatism in public organizations, and it is difficult to change existing (path-dependent) routines and practices. In addition, there is much ignorance of other and new ways of doing (for example with regard to stakeholder participation, adaptive management, and collaborative dialogues). Traditional water authorities in particular tend to stick to what they have always been doing and face many difficulties when they are asked to become more adaptive and integrative. Games and design charrettes can form an interesting platform to experiment with, and learn about, the new ways of doing (see Chapters 6, 7, 8, and 9).

Another important insight has to do with the importance of questioning the current institutional regime and analysing whether institutional redesign is necessary to enable the development and implementation of effective adaptation strategies. There is often limited 'design space' because of institutional limitations. This means that the governance of climate change adaptation should address the question not only of how to design effective measures, but also of which institutional adjustments or provisions are necessary to implement these measures.

Scientific issues: recoverability and independence of the research

Recoverability

An important quality criterion in natural and social sciences is repeatability. In Chapter 2 it was argued that action research is usually not (easily) repeatable, because it is conducted in ongoing empirical processes in complex situational contexts. The results are valid in that context, and they cannot be repeated because the situation in practice will have changed after the research. Therefore, action research can hardly be repeated elsewhere in time or space. To a certain extent, games can be repeated in other contexts, as games provide a more 'standardized' environment. However, there are limitations to this because, as Schenk and Susskind have explained, the game needs to reflect a realistic situation to a certain extent in order to be interesting and relevant enough for the participants; this means that it should not be too abstract or context free. To a certain extent, the game thus needs to be adapted to the local situation. So elements can be repeated, but there is also a certain uniqueness. Moreover, because of the learning effect for game participants, even repetition within the same group will result in different outcomes.

A quality criterion better suited to action research is recoverability (Checkland and Holwell, 1998; see also Huntjens et al., Chapter 2, this volume). Recoverability is about keeping records of the research activities so that anyone interested can subject the research to critical scrutiny (Checkland and Holwell, 1998). Can action research processes be recovered by interested outsiders? In many senses, problems of recoverability in action research do not differ from those experienced in other research approaches. Scientific guidelines to ensure transparency and provide insights into how the research was set-up and implemented can be used in action research to increase recoverability. Important steps are explicating the perspective used for analysis by explaining the theories used and neatly describing the research methodology, including how data were gathered and analysed. Keeping a reflection diary or log book (e.g. see Boezeman *et al.*, Chapter 5, this volume) can support this process. This is especially important because the course of events during an action research process is frequently adjusted consequent to unforeseen or changing circumstances.

However, the empirical chapters in this book point to several issues that deserve special attention to ensure recoverability in action research. The first issue is that the practitioners involved in action research seem to be less disciplined and less interested in following scientific guidelines to ensure recoverability. They are more interested in being accountable to their executives (and sponsors) and in the practical relevance and usability of the results than in documenting all steps to ensure scientific recoverability. Especially if action research is carried out at the co-production level (scientists and practitioners carry out the research together), the scientists will have to make sure that the different steps in the research process are well documented and carried out in a recoverable way.

A second issue regards the recoverability of interventions. As Chapters 3 and 5 have shown, both the intervention itself and the analysis of its effects can be done in a recoverable way. So there is no fundamental tension between interventions and recoverability. In the more controlled environments of games and design charrettes, it might be even easier to analyse the influence of an intervention (especially if the same intervention can be repeated in multiple sessions). Both in games and in real life, however, interventions might also have unintended consequences, or parts of the research might have consequences that were not initially intended. Boezeman *et al.* (Chapter 5) for instance, describe how a tool to map stakeholders was used later in the policy process to distinguish between primary and secondary stakeholders. This once more illustrates the need to keep good documentation of the intentions, interventions, methods used, and processes.

Moreover, the exact influence of the research on policy process is often not clear. In the Copenhagen case in Chapter 8 (Driscoll and Lehmann), the games (implicitly) advocate a combination of adaptation and mitigation, as was also implemented. It is, however, difficult to assess whether this resulted from the games played by the policymakers or whether it was a more autonomous process that coincided with their research. This seems to be a drawback of games; they increase recoverability, but the effects afterwards on policies are more difficult to monitor.

The question arises as to whether all action research methods are recoverable in the same way: does the level of interaction and involvement with the research theme influence recoverability? In principle, all action research can be carried out in recoverable ways, but with some methods and high levels of intensity it is more difficult. Action research with a lot of participatory observation means a lot of short interactions with practitioners, often in contexts where it is difficult to make field notes all the time. As argued above, the more practitioners are involved and the more intensely they are involved, the more the researcher has to do to make their activities recoverable, because practitioners themselves tend to be less interested in ensuring recoverability. It can be helpful to organize different roles in an action research team, where some people are responsible for the action component and others for the analysis and reflection component.

Independence

Action research is about entering a real-world situation and aiming to improve that situation. This implies a certain level of involvement or even engagement by the researcher. In action research on the governance of climate change adaptation, important issues include the distance from the policymaking process and the independence of the researcher on becoming involved and receiving co-financing from stakeholders. There is an undeniable tension between involvement and keeping a distance from the object of research so that one can analyse the situation without becoming biased because of personal involvement (as the more traditional scientific methodologies prescribe). However, action research on climate change adaptation shows a large variety in the degree to which researchers keep a distance and remain independent. It depends on the exact set-up of the research and the role chosen by the researcher.

The chapters involving games showed various levels of involvement. Schenk and Susskind stress the importance of creating a game outside the context of the actual decision-making process, so that the stakeholders can think and act more freely. In this way, the researcher also distances him/herself from the actual policy process. A drawback of this approach is that it becomes harder for practitioners to translate any lessons learnt back to their own situation. Both Roggema *et al.*, and Huntjens *et al.*, adopted a more involved approach in that they actively engaged in co-creating a plan with the stakeholders. Through their engagement, they might also have become committed to the outcome, and in that sense they may have lost a degree of independence. Interestingly, the chapters show that the level of involvement and commitment may vary over time: during the planning stage, the researchers were involved, but when writing their chapter afterwards they had already developed more distance from the plans. Thus, involvement and distance are not constant but dynamic, and this may be something researchers can use to realize results in practice and later reflect upon from a distance when trying to reap the scientific fruits.

Chapter 3 (Vink *et al.*) highlights how the involvement of the researcher with one of the practitioners was beneficial for the research. By linking up with a practitioner who became a 'guide' and facilitated access to the field, better data could be gathered. Other authors also emphasize the need for a high involvement of (at least a small selection of all) stakeholders in the research. Schenk and Susskind, for instance, aimed to include stakeholders in the development of the games in order to best fit the local situation. In Chapter 10, a municipal civil servant was strongly involved in the whole research process as one of the project leaders. This increased the relevance of the research for the municipality (who also co-financed the project) but also affected the independence of the researchers involved. This was (partly) resolved by making a role distinction within the research between more consultancy-oriented members and more science-oriented members. The same holds true for the Ellen *et al.* case, in which a policy professional from the provincial government participated.

Co-financing is often applied to make action research projects possible. The project is then co-financed by one or more stakeholders who want a particular problem to be solved. In the chapters by Ellen *et al.* and van Buuren *et al.*, the co-financers thus enabled action research and collaborated with scientists in a productive way but also influenced which problem was addressed by the researchers. Co-financing thus has a paradoxical consequence. On the one hand, it enables the research as such to be conducted and opens the door to knowledge development. It also guarantees the policymakers' collaboration, which is important for successful action research. On the other hand, it gives policymakers also the role of principal, and this can negatively influence the 'research design space' that researchers obtain to optimize the study.

It is not only the distance from the policy field in general that is important, but also the distance that the researcher keeps from particular interests or points of view. To what extent does the action researcher align him/herself with particular parties, or with particular perceptions of the problem and desirable solutions? In other words, to what extent does the researcher take a normative stance? We elaborate on this in the next section.

Normative issues

In this book, climate change is approached as a governance issue, and the researchers have involved themselves through action research. This position does have normative implications. First of all, it implies that climate change is considered something to be dealt with; it requires action and should not be neglected. In addition, working on climate change *adaptation* – mainly dealing with the consequences of climate change – often also implied less attention for strategies designed to fight the causes of climate change (mitigation, although there can also be synergies between them, see Driscoll and Lehmann, Chapter 8). Further, in several chapters, adaptation to climate change is also framed as a *governance* problem; this implies that it cannot be solved by one actor alone. The problem is implicitly or explicitly framed as a problem of joint action and cooperation.

Underneath these general characteristics, the chapters show differences in the degree to which the researchers aligned themselves with one particular solution or interest. In most cases, the researchers aimed to facilitate the process of multiple actors arriving at a commonly accepted solution, without taking a substantive position as to which solution would be the best one. In this sense, their approach was different from the tradition of participatory action research, which often has an emancipative goal and takes sides to improve the position of certain disadvantaged groups who have little power in climate change projects (e.g. citizens, farmers). However, in most cases, (social) learning by the involved stakeholders was at least an implicit goal.

One could argue that, in contested policy fields, not taking a particular substantive position is also a way of taking a position, because by not supporting disadvantaged groups one reproduces the dominance of vested interests and existing policy programmes. In other words, there is no way of avoiding a normative position in a contested policy field such as climate change. As stated, in most chapters, the researchers tried to facilitate the process of arriving at consensus about measures for climate change adaptation. In some cases, they actively took a position to support a particular plan, including the substantive choices. In other cases, they tried to avoid a normative substantive position. Vink *et al.* took a neutral stance, but their explication of frames other than the dominant frame was the start of the process in which the official frame was changed to incorporate aspects of other frames as well. Thus, their research was not neutral in its effects. Likewise, Boezeman *et al.* aimed to take a neutral stance, but in the end their action research did influence the governance process and gave water board officials a tool to differentiate between important and less important stakeholders. Even when researchers aim to take a neutral position, their research is not always neutral in its effects.

Even though taking a certain position can hardly be avoided, action researchers may still take the role of explicating and reflecting on the various normative positions that stakeholders adopt. One example of this is the research of Vink *et al.* who explicated the frames used by a crucial governmental stakeholder. They revealed the normative frame and thus gave the involved actors the possibility to take this into account and reflect on it.

The practical value of action research

The added value of action research for climate change adaptation practices

From this book, we can distil at least four functions of action research in climate change adaptation. First of all, action research helps to unlock scientific knowledge for practitioners and policymakers. For many policymakers, it is a welcome opportunity to get up-to-date scientific information, empirical insights from other cases or even other countries. This is clearly shown in Chapters 5 and 7. Scientists involved in action research can be seen as the gateway for practitioners to become informed about what is happening in relevant scientific areas like planning, environmental studies, public administration with regard to issues of participation, policy instruments, management paradigms, and so forth. Action researchers are thus often public officials' informal knowledge brokers.

Second, action research actually contributes to the organization of collaborative governance processes and the bringing together of relevant actors. Some chapters (e.g. Chapters 4, 5, 6, and 10) describe how scientists assist policymakers by organizing collaboration with societal stakeholders and/

or citizens. For stakeholders, the involvement of scientists can contribute to their ease-of-mind and thus their willingness to participate. For policymakers, the involvement of scientists is a welcome source of assistance. This is useful for them in successfully organizing stakeholder participation, especially when they do not have experience of that.

Third, collaborative action research enhances reflection and learning among practitioners (policymakers, stakeholders). This is true for the various games included in this volume, but also for the other cases (see e.g. Vink *et al.* who aimed for frame reflection and Roggema *et al.* who switched participants' roles to stimulate learning and reflection). Interaction with scientists helps stakeholders to reflect upon their own practices and facilitates processes of single- and double-loop learning, for instance during collaborative governance processes.

Finally, action research in the context of climate change adaptation helps policymakers to develop concrete arrangements and strategies to cope with it. In only a minority of the cases was action research used to co-construct strategies and arrangements, but it is certainly a possible function of action research. It requires scientists to be able to translate their scientific insights into applicable notions about how to organize adaptation strategies. It also presupposes that policymakers are willing to become involved in a process of experimentation, fine-tuning, and reflection in order to enable validation and deduction. This is rather difficult, and scientists have to be careful not to become consultants with little or no opportunity to safeguard the scientific quality of their data. There is still a long way to go to enhance the intervention and design component within disciplines like planning studies, governance studies, and public administration sciences.

Practical issues: timing, resources, urgency

There are at least three major practical issues that complicate the organization of effective action research projects. First of all, there is the tension of synchronizing different time orientations and perspectives. Whereas policymakers attune their actions to the ongoing dynamics in political processes, researchers are focused upon running a linear process from theory development, to empirical research, analysis, and reporting. These different time logics and perspectives are difficult to match. They require both sides to be willing to be flexible, but, most of the time, administrative and political agendas will dominate, and researchers have to adjust their planning and project to what the policymakers are doing.

Second, there is the issue of resources. Action research is both time-consuming and labour-intensive. With shrinking budgets and a growing workload, the availability of sufficient resources (money, attention, support, and time) is, however, problematic for both policymakers and researchers. Many case studies concluded that their results would have been better if there had been more time and money for conducting the research.

A third issue has to do with mental factors like urgency, commitment, mutual trust, and reciprocity. With their many conflicting demands, it is difficult for policymakers to commit themselves to processes of intense collaboration with scientists. In addition, climate change adaptation is currently often not very high upon the internal political agenda. Therefore, it is difficult to convince policymakers of the relevance of conducting action research. Scientists have to dispel various prejudices against them with regard to the relevance and usefulness of their knowledge (e.g. they approach a problem in too difficult/abstract a manner and it is therefore hard to translate their findings into knowledge that is readily useable and relevant for the policymaker).

Action research as co-production of application-oriented knowledge

All in all, action research is aimed at co-producing knowledge that is not only usable and policy relevant but also scientifically valid. This co-production is far from easy. All collaborative processes face challenges in relation to conflicting perceptions, language, time horizons, and task orientations.

Fortunately, in the worlds of both science and policy, we can see tendencies that potentially contribute to better cooperation. In the policy domain, the room for experimentation, gaming, and pilots seems to be growing, as is also illustrated by several chapters in this book. Vink *et al.* were allowed to observe and reflect on the policy process in the Lake IJssel region. Likewise, Driscoll and Lehmann observed policymakers in Copenhagen, who also participated in their serious games. Roggema *et al.* were allowed to engage closely in the design of spatial plans. At the same time, we can see a growing attention on the valorization and utilization of scientific knowledge; this development forms a strong driver for scientists to strengthen their orientation towards policy practice.

Ultimately, science also has a critical ambition. It has to provide society with a critical perspective, unravelling the weaknesses in existing policy paradigms and inconsistencies in public discourses. Working together with policymakers makes it difficult for researchers to remain independent and critical. On the other hand, it increases the chances of actually reaching and influencing policymakers. Dealing with this dilemma is not definitively solved in this book. One way in which many scientists try to deal with this tension is by placing different emphases on stories for different publics. They are more critical in their scientific papers and more conciliatory in their policy advices.

Conclusions and discussion

Action research serves as a bridge for scientists and policymakers to meet each other and to become better informed about what is happening in 'the other world'. This is important, as both hold valuable knowledge and insights. A first function of action research is thus to enhance knowledge dissemination and the valorization of scientific knowledge. Related to this function, action research contributes to reflection and learning amongst policymakers. It can enhance awareness about climate change and the need for adaptation, and can offer insights into alternative strategies and possible instruments.

At the same time, conducting action research in the field of climate change adaptation can also start from purely scientific aims, for instance improving our understanding of decision-makers' behaviour and implicit action frames, or studying which governance instruments and approaches might be most effective for delivering adaptation measures.

Action research thus can enhance our understanding of complex governance processes. Their close involvement in practice provides scientists with insights that they might not have gained via traditional approaches. Such empirical data can lead to new ideas on how the governance of adaptation to climate change functions. These results can also be directly shared and discussed with policymakers. This can help to test the credibility and usability of the results for policymakers.

However, it must be admitted that conducting action research is far from easy. This chapter has addressed a number of serious pitfalls that need attention when action research is being conducted in the field of governance of adaptation: two of them are *recoverability* and *independence*. Recoverability is important in order to give people interested in the research a better idea of how it was conducted and how results were obtained. Recoverability is an issue because action research methods influence the process under study. This might be intentional (for instance for normative reasons) but can also be unintentional (e.g. the impact of a stakeholder analysis on the rest of the decision-making process and its outcomes, see Boezeman *et al.*). Unintentional influences in particular might only be transparent when good notes are taken so that the process and the researcher's choices are recoverable.

To safeguard their independence, researchers need to keep enough distance from their object of study. At the same time, action research requires close cooperation with practitioners and involving them in the research design and execution. The researcher should also incorporate moments during action research to step backwards and reflect upon the course of events. This tension further underlines the need for good reflection/research diaries. It can also be helpful to organize different roles in a research team (with some researchers being at a distance and others more deeply involved in the actual action research).

From this volume, we can learn that co-production of actionable knowledge is a balancing act, in which actors have to balance between sufficient distance (in order to prevent blurring boundaries with bad science and bad policy as a result) and enough proximity (to facilitate direct interaction and joint sense-making). In the case of 'embedded science', there are serious issues to address regarding recoverability and validity. We hope that this volume can contribute to dealing with these issues, especially in the new and dynamic field of the governance of adaptation to climate change. This field is beset by many complex challenges, and there is a strong need for application-oriented knowledge, as well as for critical policy advice on how to govern climate change adaptation and deal with the specific challenges it poses. We hope that this volume has shown that action research can indeed yield valuable results that improve both the governance of climate change adaptation and scientific theories.

References

Berkhout, F., G. Verbong, A.J. Wieczorek, R. Raven, L. Lebel and X. Bai (2010) Sustainability experiments in Asia: innovations shaping alternative development pathways? *Environmental Science & Policy*, 13(4): 261–271.

Buuren, M.W. van, P.P.J. Driessen, G.R. Teisman and H.F.M.W. van Rijswick (2014) Toward legitimate governance strategies for climate adaptation in the Netherlands: combining insights from a legal, planning, and network perspective, *Regional Environmental Change*, 14(3): 1021–1033.

Checkland, P. and S. Holwell (1998) Action research: its nature and validity, *Systemic Practice and Action Research*, 11(1): 9–21.

McFadgen, B. and D. Huitema (under review) Experimentation and learning, the design of policy experiments and their learning effects: a conceptual framework and application to a case study from the Netherlands, *Ecology and Society*.

Index